THE
FAT KITCHEN

THE FAT KITCHEN

HOW TO
RENDER, CURE & COOK
with Lard, Tallow
& Poultry Fat

ANDREA CHESMAN

Storey Publishing

The mission of Storey Publishing is to serve our customers by publishing practical information that encourages personal independence in harmony with the environment.

Edited by Carleen Madigan
Art direction and book design by Alethea Morrison
Text production by Jennifer Jepson Smith
Indexed by Andrea Chesman

Cover and interior photography by © Keller + Keller Photography
Additional photography courtesy of Andrea Chesman, xii; © Apic/RETIRED/Getty Images, 13; © Glow Images, Inc./Getty Images, 20; © INTERFOTO/Alamy Stock Photo, 181; © Jeff Morgan 10/Alamy Stock Photo, 2; © Ronald C. Modra/Sports Imagery/Getty Images, 31
Food and photo styling by Catrine Kelty
Lettering design on cover by © Tobias Saul
Lettering design pages xiv and 94 by © Kateryna Dekhtiarenko

Be sure to read all of the instructions thoroughly before undertaking any of the techniques or recipes in this book and follow all of the recommended guidelines.

Storey Publishing
210 MASS MoCA Way
North Adams, MA 01247
storey.com

Printed in China by Toppan Leefung Printing Ltd.
10 9 8 7 6 5 4 3 2 1

LIBRARY OF CONGRESS CATALOGING-IN-PUBLICATION DATA

Names: Chesman, Andrea, author.
Title: The fat kitchen : how to render, cure & cook with lard, tallow & poultry fat / Andrea Chesman.
Description: North Adams, MA : Storey Publishing, [2018] | Includes bibliographical references and index.
Identifiers: LCCN 2018021356 (print) | LCCN 2018022130 (ebook) | ISBN 9781612129143 (ebook) | ISBN 9781612129136 (pbk. : alk. paper)
Subjects: LCSH: Oils and fats, Edible. | High-protein diet—Recipes. | LCGFT: Cookbooks.
Classification: LCC TX407.O34 (ebook) | LCC TX407.O34 C44 2018 (print) | DDC 641.5/638—dc23
LC record available at https://lccn.loc.gov/2018021356

Contents

Foreword by Michael Ruhlman vii
Preface ix

UNDERSTANDING ANIMAL FATS

RECIPES

Foreword

This is a book after my own heart. Unapologetic, informative, clear, and filled with recipes that make me want to head to the kitchen. Roast duck and southern-style green beans (appropriately cooked forever), red bean gumbo and pasta e fagioli, scallion pancakes and bacon jam and Korean fried chicken wings, a blueberry galette and a simple apple tart. Such compelling variety, and all inspired by fat, America's bugaboo. Until now, I hope.

Andrea Chesman has created an exceptional cookbook. The first third of it is all about fat — its chemistry, the different types, the good and the bad of fats, and how they work in your body. I loved her straightforward explanations, which are remarkable for their objectivity and balance. She tells you what's what — you decide if you want those duck fat fries (short answer: you do). As I like to say, fat isn't bad; *stupid* is bad. Healthful cooking and eating is about balance, not avoidance.

We live in a time of widespread confusion about food. And it's no wonder, given our current obesity and diabetes epidemics. But these ills are not likely because of the fat we eat; as research is beginning to show, sugar and refined wheat seem to be. We need more clear voices in the food world, those who speak from the heart and create books such as the one you're holding.

The Fat Kitchen is an embrace of life and living and the great, deep pleasures — not of fat, but rather of cooking well, with love and with respect, hearing the deep whispered joy of a good meal shared with the people you love.

— **MICHAEL RUHLMAN**, author of 17 books, including
Charcuterie, The Book of Schmaltz, and
Grocery: The Buying and Selling of Food in America

Preface

Some of us come to cooking with animal fats out of a desire for better health, some for thrift, some for the flavor, some for better baked goods. Chances are, if you're picking up this book for one reason or the other, you will stay for all of the above reasons. I have.

When I was very young, my grandmother, a Jewish immigrant from Warsaw, lived with us. She ruled the kitchen, and everything she prepared started with a scoop of chicken fat. In a similar way, her non-Jewish compatriots — from Poland and throughout Eastern Europe — started their recipes with a spoonful of lard.

Fat tastes good — and it defines the flavor of traditional cuisines around the world: Think duck fat and butter in France, ghee in India, olive oil in Italy, lard in Mexico, schmaltz (rendered chicken fat) in Jewish cuisine. Salt pork was crucial in defining the flavors of American cooking, especially in New England and the American South, where they called it "streak of lean." In England, cooks made use of beef fat in the signature dish of Yorkshire pudding (made with beef drippings) and in many pies and pasties (made with tallow in the crust). Suet — unrendered beef fat — is a necessary ingredient in many English steamed puddings.

Lard and other fats derived from pigs were once the most widely used fats all around the world. Even in Italy, lard was used extensively, especially in the north, where the climate did not lend itself to the growing of olives. Dishes were often started with a battuto, a combination of finely chopped onions, celery, carrots, garlic, and parsley cooked in lard. In Spanish households, that same combination was called a sofrito; in France, a mirepoix.

But with the rise of industrialized food production the world over, fats derived from animals fell out of favor. Some of this was due to the cheap production of vegetable seed oils, helped along by some very clever marketing by Crisco and Wesson, which touted the "purity" of their products. At the same time, the urbanization of the population and a change in how meat was raised and purchased in supermarkets helped changed the central role animal fats once played. But much of the switch from animal fats to industrial vegetable seed–based fats and oils was due to flawed medical science, which drew a link between the consumption of animal fats and heart disease. An army of public health advocates and government policy makers spread that false claim like it was gospel.

All animal fats are not created equal because all animals are not raised under the same conditions. When I write about cooking with animal fats, I am writing about fats from pasture-raised chickens, ducks, geese, pigs, lamb, goats, and cattle. I am not writing about animals raised or finished on grain in concentrated animal feeding operations (CAFOs). Such large-scale animal operations aren't remotely healthy environments for the animals, so the animals are dosed with antibiotics, which also allows them to gain weight faster. This, in turn, alters the flavor and nutritional content of the fat. The fat from industrial-farmed animals contain fewer anti-inflammatory omega-3 fatty acids and more pro-inflammatory omega-6 fatty acids than fat from pasture-raised animals.

In the years that I have been thinking and learning about animal fats in the diet and writing this book, animal fats have started to receive more and more approval from the scientific community, chefs, and consumers. Friends and relatives no longer respond to the idea of this book with disapproval and dismay. The continuing popularity of the paleo diet and the work of organizations devoted to traditional foodways, such as the Weston A. Price Foundation, has generated a groundswell of interest in the benefits of cooking with animal fats instead of vegetable seed oils. In early 2017, Sally Fallon Morell came out with *Nourishing Fats: Why We Need Animal Fats for Health and Happiness*. *The Big Fat Surprise* (first published in 2014) by *New York Times* reporter Nina Teicholz has provided well-documented proof that the connection between saturated fats and heart disease is just not evidence based. Mark Hyman of the Cleveland Clinic Center for Functional Medicine has written convincingly about the role of "good fats" in the diet, and some health professionals have begun to dial back on their blanket condemnation of animal fats. Meanwhile, a lot of studies point to an excess of sugars and carbohydrates as the real public health menace, a point the sugar industry has worked hard to suppress.

Some health professionals have begun to dial back on their blanket condemnation of animal fats.

Famed chef and restaurateur April Bloomfield, author of *A Girl and Her Pig* and other cookbooks, opened a butcher shop in Manhattan in 2015 that is also a casual spot for breakfast and lunch and a more formal restaurant at night. She named it White Gold, a reference to animal fat — and she makes great use of it, in potato pasties that are made with chicken fat, in Hasselback potatoes that are deep-fried in beef tallow, in radishes that are sautéed in marrow butter, and in countless other dishes.

Many other restaurants are boasting about the animal fats they use, especially duck fat for French fries. Indeed, there is a restaurant in Portland, Maine, by the name of Duckfat. It doesn't include duck fat in every dish on the menu, but it does offer a few. The restaurant was an early adopter of cooking with delicious duck fat in this country, but many more restaurants are using it with pride. Today there are plenty of upscale burger "bistros" that offer French fries cooked in duck fat. And, since duck fat is a such a tasty addition to most foods, trendy menus are offering snacks like popcorn and even Chex Mix with duck fat upgrades. Once found only in France — or in restaurants or in homes where the cooks rendered the fat from the ducks they bought — today duck fat is more widely available online. (For a listing of online sources of animal fats, see Resources, page 278.)

Although by now all of us should be well informed that there is no connection between the consumption of animal fats and heart disease, many still believe a connection exists. After all, it has been preached and promoted by the medical establishment and the U.S. government for about 60 years. People still reach for low-fat processed foods, even though such foods may be loaded with sugar to compensate for the flavor lost when fat was eliminated, and mainstream medical professionals are still advising people to cut back on fats and to prefer unsaturated fats over saturated fats. Is that wise? Is it borne out by the latest research?

In this book, I cover why animal fats belong in a well-balanced diet, and how to source high-quality ones. I cover how to render fats, how to store them, and how to cook with them. I am not including much about butter in this book because there is a wealth of information out there about cooking and baking with butter. You have undoubtedly cooked and baked with butter yourself, or others have prepared food with butter for you. But have you tasted a superior piecrust or a biscuit made with lard or beef tallow? Have you tasted what duck fat can do for roasted root vegetables? For potatoes prepared every

My grandmother Esther Lewin lived with my family when I was young. She started most dishes by sautéing onions in schmaltz.

which way? Did you know that frying in lard or tallow leaves the finished food crispier and less greasy than when the same food is fried in oil at the same temperature?

There are tricks to working with animal fats — ones your grandmother or great-grandmother knew — but they haven't been passed on. Indeed, just about all that remains of that traditional knowledge is how to make a piecrust with half lard and half butter and how to throw some salt pork into Southern-style vegetables and Boston baked beans for flavor.

This book also contains recipes; most of them are traditional recipes from around the world. Although you can adapt just about any recipe to using an animal fat instead of oil or butter (and I cover that, too), for the recipes gathered here, I have focused most of my attention on the dishes that have evolved over time in cultures that have used animal fats freely. A cookie that was traditionally made with lard doesn't taste right with butter. Chicken fried in lard has far better texture than chicken fried in oil. Here you'll find many dishes that are just plain better in terms of taste and texture when made with animal fats, and that includes almost all potato dishes — and popcorn. It also, surprisingly, includes blueberry muffins.

By cooking with animal fats from pasture-raised animals rather than from industrialized vegetable seed oils (corn oil, soybean oil, rapeseed oil, and so on), you will also be promoting grassland biodiversity and carbon sequestration. You will be reducing consumer demand for palm kernel oil (which is heavily used in processed foods) and coconut oil, and thereby refusing to be part of the system that is devastating the rain forest in equatorial areas. You will be allowing grass farmers to reap the full value of the animals they raise by making the fats a saleable commodity. How can all that not be good for us?

CHAPTER 1

A Little Chemistry, A Little Biology

*T*he word "fat" has many meanings. It can be a disparaging adjective to describe an overweight person; it can be an approving adjective, as in describing a well-supplied bank account. "Fat" is also a noun, a substance found in foods and in our bodies. It can be invisible, as in, say, avocados, or visible, as in a well-marbled steak. It can be a cooking medium, as in a liquid like corn oil or a solid like lard.

We associate fat — the cooking medium and the fat in foods — with the fat that accumulates in our bodies. This has been an association that has been drummed into our heads by the medical and dietary establishment for more than 60 years. It seems so obvious, no?

No.

It is time to take a peek behind the science of healthy eating.

THE CHEMISTRY OF FATS

To a chemist, all edible fats and oils are lipids, and they are all basically the same, even though in the kitchen we often say oils are liquids and fats are solids at room temperature. The simplest definition is that all fats and oils — all lipids — are biological chemicals that do not dissolve in water, which is why they rise to the top of liquids, allowing us, for example, to skim the fat off the tops of soups and stews, if we want to.

On the molecular level, these biological chemicals are chains of fatty acids, divided into two broad categories: saturated fatty acids and unsaturated fatty acids. Saturated fatty acids (the building blocks of saturated fat) are chains of carbon atoms with pairs of hydrogen atoms joined at each link with a single bond. They are *saturated* with hydrogen atoms, leaving them no room to bond with other atoms. This makes them stable and allows the fatty acids to pack together tightly, which causes them to be solid at room temperature. Unsaturated fatty acids (the building blocks of unsaturated fat), on the other hand, are chains of carbon atoms that have some double bonds between links with gaps where the hydrogen atoms are missing. If one hydrogen atom is missing, the fat is monounsaturated; if more than one hydrogen atom is missing, the fat is polyunsaturated. These fatty acid chains are *unsaturated* with hydrogen atoms and contain open bonds that easily bind with oxygen. Monounsaturated fats, like olive oil, have only one open link; polyunsaturated fats, like corn oil, have more than one open link.

Unsaturated fats are liquid at room temperature and readily bind with oxygen, because of those open bonds. When exposed to heat or air, an unsaturated fat oxidizes, allowing oxygen molecules to bind to its open bonds. A fully oxidized fat is rancid. The more available a fatty acid is to binding with oxygen, the easier it is for it to become rancid.

RIGHT: Because of their molecular structures, vegetable seed oils (like the olive oil pictured here) are liquid at room temperature, while lard is solid.

No Fat Is Totally Saturated

The idea that animal fats are all saturated fats, that olive oil is all monounsaturated fat, and that vegetable seed oils are completely devoid of saturated fat is a gross oversimplification. The truth is that many animal fats have a high percentage of monounsaturated fat, bringing them close to olive oil in terms of fatty acid content. Butter is about 63 percent saturated fat, beef tallow contains about 41 percent saturated fat, lard is about 39 percent saturated fat, and chicken fat is about 30 percent saturated fat. So when we say that animal fats are made up of saturated fat, it is important to understand that each animal fat is made up of some saturated and some unsaturated fat, and most of that unsaturated fat is the so-called good kind — monounsaturated fat. The proportion of saturated and unsaturated fats varies among the different fats and oils.

As you can see from the graphic on page 7, lard, poultry fats, and beef fat are all relatively high in monounsaturated fat. It is the monounsaturated fat that health professionals agree is the healthiest to consume, because they can help reduce the so-called bad cholesterol (LDL) levels (more on that later) in your blood, and thereby lower your risk of heart disease and stroke. So if you look at corn oil, which is touted as healthy because it is high in polyunsaturated fat, note that it is relatively low in monounsaturated fat, compared to beef tallow, lard, or any poultry fat.

By the way, much of the monounsaturated fat found in animal fats is oleic fatty acid, which is also found in olive oil and associated with decreasing LDL cholesterol, thus lowering "bad" cholesterol.

How an Oil Is Processed Matters

Canola oil and olive oil are the two oils that have the highest percentage of monounsaturated fat, but where olive oil is often cold pressed (as virgin or extra-virgin olive oil), canola oil is chemically extracted from rapeseeds, then bleached and deodorized. In the process, it is subjected to high heat, which destroys the vitamins and any antioxidants it may have originally contained. In addition, 90 percent of the rapeseed crop (the seed from which canola oil is

COMPARING FATS

% **Saturated Fat**

% **Monounsaturated Fat**

% **Polyunsaturated Fat**

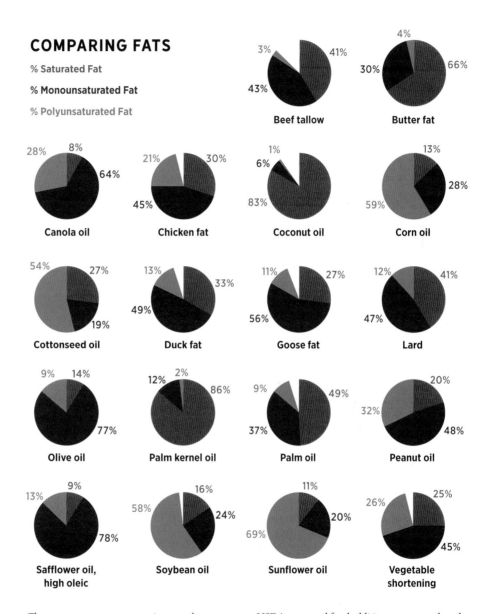

Beef tallow — 3%, 41%, 43%

Butter fat — 4%, 66%, 30%

Canola oil — 28%, 8%, 64%

Chicken fat — 21%, 30%, 45%

Coconut oil — 1%, 6%, 83%

Corn oil — 13%, 28%, 59%

Cottonseed oil — 54%, 27%, 19%

Duck fat — 13%, 33%, 49%

Goose fat — 11%, 27%, 56%

Lard — 12%, 41%, 47%

Olive oil — 9%, 14%, 77%

Palm kernel oil — 12%, 2%, 86%

Palm oil — 9%, 49%, 37%

Peanut oil — 20%, 48%, 32%

Safflower oil, high oleic — 13%, 9%, 78%

Soybean oil — 16%, 24%, 58%

Sunflower oil — 11%, 20%, 69%

Vegetable shortening — 26%, 25%, 45%

These percentages are approximate; values vary depending on the source. Animal fats contain varying amounts of water and connective tissue. Manufacturers may add low levels of FDA- and USDA-approved food additives to protect the oils during processing, storage, handling, and shipping. Because of these additional ingredients in both oils and fats, the numbers may not add up to 100%.

extracted) is grown from genetically modified seeds. To my mind, this calls into question the whole idea that canola oil is a healthy choice — for the planet, if not for ourselves.

To be fair, I note that you can buy canola oil that is cold pressed from organically grown, non-GMO (genetically modified organism) seeds, but it is quite expensive compared to the standard supermarket brands. Also, whereas the smoke point of refined canola oil is 400°F, which makes it suitable for deep-frying and other high-heat applications, the smoke point of unrefined canola oil is 225°F, making it suitable only for salad dressings or other uncooked preparations.

Unless a vegetable seed oil is cold pressed, it goes through a process where a solvent such as hexane is used to maximize oil extraction. The refined oil is then mixed with a powdered bleaching agent, which binds to the unwanted substances left in the oil, and then is filtered out. At this point, the oil has a

NOT ALL TRANS FATS ARE EQUAL

Industrially made trans fats, produced by adding hydrogen molecules to vegetable seed oil and lard, have been added to all sorts of foods to increase their shelf life. Such man-made trans fats increase the risk for heart disease, and, as of 2018, have been banned from the U.S. food supply.

Naturally occurring trans fats are created during the digestion process of the ruminant animals (cattle, sheep, goats). So you may see trans fats listed on the labels of some dairy products and tallow, especially tallow made from grass-fed cattle. Oddly enough, only tallow from grass-fed cattle will have enough trans fats to register on the labels required by the United States Department of Agriculture (USDA) for foods containing above 0.5 gram per serving. These naturally occurring trans fats are not added to tallow or dairy products and don't seem to have any negative health impacts — they may even be protective of heart health.

disagreeable odor, so it must be further processed. To deodorize, the oil is heated to a high temperature and pretty much any remaining components other than fat are cooked off (including antioxidants). This refining process leaves edible oils devoid of the majority of their natural nutrients and damaged due to exposure to high heat. Exposure to high heat makes all fats and oils prone to becoming rancid — whether you can taste it or not.

If It's Rancid, Throw It Out!

Remember those unstable bonds in polyunsaturated fats? As the fatty acids bind to oxygen, the fatty acids break down, becoming rancid. A significantly rancid fat has a distinctive flavor and odor, which some describe as paintlike or grassy. It smells like turpentine to me. And it isn't only oils that become rancid. Rancidity happens in grains and nuts, too. Think about tasting a cracker from a box that has been sitting on the shelf for too long. That "stale" taste is rancidity. Rancid odors and flavors are found throughout the food chain, and one concern is that Americans are so used to eating rancid foods that either they don't recognize the flavor of rancidity or they eat rancid foods anyway out of a misplaced sense of thrift.

All fats can become rancid. Any fatty acid chain that is unsaturated can become oxidized, meaning the carbon-carbon double bond in the structure can be broken by oxygen in the air, forming a carbon-oxygen bond. When this carbon-oxygen bond breaks down, aldehydes, ketones, or carboxylic acids are formed; these are the compounds that produce rancid odors and flavors. Rates of oxidation are sped up in the presence of light and heat, which is why you should always store cooking oils in tinted bottles in a dark cupboard with the cap secured and store cooking fats in the refrigerator.

The more polyunsaturated fat an oil contains, the faster it can become rancid. Flaxseed oil, sunflower oil, and corn oil will become rancid faster than other oils. The potential for animal fats to become rancid is much less, because of their lower percentage of polyunsaturated fat. That large, clear, plastic bottle of corn oil sitting on a kitchen counter near the stove has ample time to be exposed to light and heat and become rancid. So, too, does that canning jar of saved bacon fat that is added to and subtracted from over the years (better to keep it in the fridge). Note that in the olden days, bacon fat was likely stored in an old tin can, which at least blocked the light.

What's so bad about rancid fats? Rancidity reduces the vitamin content the fats and oils may have had originally. More dangerously, rancid fats and oils can develop potentially toxic compounds that have been linked to inflammation, which has been linked to advanced aging, neurological disorders, heart disease, and cancer. Inflammation is also linked to diabetes, arthritis, and countless immune system disorders. Given how many potentially toxic compounds we are exposed to in our environment, it makes sense to limit rancid fats in our foods.

The shelf life of vegetable oils high in polyunsaturated fat is about 6 months in an unopened bottle, and 1 to 3 months once the bottle is opened. Some specialty oils, such as sesame, walnut, and flaxseed, have shorter usable lives. That does not stop supermarkets and big box stores from offering gallon jugs of oil for cooking.

The shelf life of home-rendered poultry fats (chicken, geese, duck) is about 4 months in the refrigerator and much longer in the freezer. The shelf life of beef and pork fats is about 12 months in the refrigerator and much longer in the freezer. Like seed oils, animal fats will not go moldy (unless dirty spoons are used for scooping), but they will become rancid over time.

Here's the thing to remember about rancidity: It is a process. It is not like your foods are not rancid one day, and rancid the next. They are always in a state of becoming rancid, but you just might not detect it.

> *Rancid fats and oils can develop potentially toxic compounds that have been linked to inflammation, advanced aging, neurological disorders, heart disease, and cancer.*

I've read a lot of blog posts about people who keep their fats at room temperature and never notice a rancid smell. I don't want to be a scaremonger, but surely a better idea would be to store the fat in the refrigerator, or freezer, or at least in a cool root cellar, where it is exposed to less heat and light. I've also read about people canning their rendered fat in a pressure canner or hot-water bath. Besides the fact that the USDA has not developed any safe times for home canning of animal fats, it seems unwise to further expose the fat to heat, thus encouraging more rancidity.

By the way, when vegetable seed oils are heated, there is a release of high concentrations of aldehydes, which is the same as what happens when oils become rancid. Heated sunflower oil and corn oil release into the air aldehydes at levels 20 times higher than the limit recommended by the World Health Organization. So repeated frying in any vegetable oil means the food and air venting from the fryer is increasingly contaminated with aldehydes, which are linked to causing cancer, dementia, and inflammatory diseases. The best restaurants change their frying oil every day, but many smaller mom-and-pop eateries — and people who own electric deep fryers — do not.

ANIMAL FATS AND HEALTH

I'm not a scientist or a health professional. I am not recommending a high-fat diet. I am not recommending that you enjoy Duck Fat Maple-Caramel Popcorn (page 100), Lithuanian Bacon Rolls (page 221), or Cheddar Cheese Crackers (page 102) on a daily basis. I am not recommending a steady diet of Hand-Cut Fries (page 203), Spicy and Extra-Crunchy Southern Fried Chicken (page 147), and Spaghetti alla Carbonara (page 141). Nor do I think every dinner deserves dessert, though I have a chapter filled with them. But if you are making a pie anyway, why not use lard or tallow instead of shortening, or instead of buying a refrigerated crust? And if you are making a healthy stir-fry, why not replace the peanut oil with lard, tallow, or a poultry fat?

Evidence has been accumulating for years that the recent shift from animal fats to polyunsaturated vegetable oils has not been healthy for us (see Resources, page 278). While some of the health problems that have arisen can be attributed to the way low-fat diets inevitably become high-carb diets, there is a good deal of evidence that industrially produced vegetable seed oils are a poor dietary choice. But that information has not yet been widely disseminated. Indeed, the dietary guidelines released by the federal government in 2015 still encouraged the consumption of polyunsaturated fats.

People have been eating animal fats for millions of years, as a source of calories and flavor; we have evolved eating them. When medical researchers noticed a rise in heart disease in America in the twentieth century, they looked to diets for the cause. They didn't look for a relationship between processed foods and heart disease. The researchers were specifically looking for a positive correlation between meat consumption and heart disease, so they found it. The idea that eating fatty meat leads to high cholesterol levels, which leads to heart disease, was so appealingly logical that studies that contradicted this neat little theory were disregarded.

The rise in heart disease rates also paralleled another major shift in the diet: the increased consumption of vegetable seed oils made from corn, soybeans, rapeseed (canola), and cottonseed. Up until this point, the only oil used in traditional diets was olive oil. Seed oils were used for making soap, but not for cooking because they became rancid too quickly or they didn't taste good. Corn oil wasn't even invented until 1908; canola oil didn't hit U.S. markets until 1978. Compared to animal fats, vegetable seed oils are a very, very recent introduction to our diet — and their health impact was not studied before their introductions.

THE "FRENCH PARADOX"

In the late 1980s, an epidemiological study revealed that the French, particularly those in southwestern France and Provence, had low rates of coronary heart disease despite a diet high in saturated fats and cholesterol. It was dubbed the "French paradox," and researchers attempted to explain it away with theories about the heart-protective qualities of red wine and a diet rich in fruits and vegetables. How about the idea that duck fat, goose fat, and lard are high in both saturated fat *and* monounsaturated fat and are not bad for your health?

Enter Crisco

In 1911, Procter & Gamble solved the problem of keeping cottonseed oil from becoming rancid at room temperature by hydrogenating the oil, or adding hydrogen molecules to the cottonseed fatty acids. Heavily promoted as a purer, healthier, more convenient product than lard, Crisco was quickly adopted because it was an alternative to an unregulated meat industry that made a regular practice of adulterating lard, often with cottonseed oil. Crisco also replaced lard and butter in a lot of baked goods and processed foods because it allowed packaged baked goods to remain fresher longer (think of the Twinkie). Was it healthier than lard as it was billed? No, not at all.

Hydrogenating oil, it turns out, creates artificial trans fats, which work to lower the so-called "good" (HDL) cholesterol while increasing the "bad" (LDL) cholesterol in the blood. In fact, trans fats are so unhealthy, manufacturers were ordered to put the amount of trans fats a food contains on labels in 2004, then phase it out from manufactured foods entirely as of 2018.

End of problem? Not quite.

Heart disease, in many cases, is caused by atherosclerosis, a buildup of plaque that makes the arteries narrower, so less blood can flow through. It seemed intuitively obvious to researchers in the 1930s and 1940s that if a person cut back on eating foods high in fats and cholesterol, the risk of heart disease would be reduced. Researchers set out to prove this "lipid theory."

Was Crisco healthier than lard as it was billed? No, not at all.

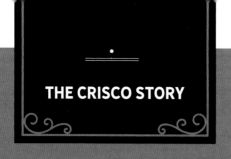

THE CRISCO STORY

When a candlemaker from England by the name of William Procter and a soapmaker from Ireland named James Gamble married a pair of sisters, the two brothers-in-law started a company that would change the diet of Americans for generations. Procter & Gamble decided on a daring scheme: to mass-produce soap as individual bars, each in its own wrapper.

Since prehistoric times, soap had been made with animal fats boiled with ashes. When soapmaking moved out of the home and into the factory, it was manufactured in large wheels, and merchants would slice off wedges for retail sale. Procter & Gamble's individually wrapped, pure white Ivory soap, which floated because it contained a lot of air, was a game changer for the industry. Plus, the marketing for Ivory soap focused on the purity of the product and convinced consumers they didn't want to have to fish around in the bottom of a tub to find the soap. Ivory soap became wildly popular.

But the nimble company needed a less expensive fat than lard or beef tallow for their soap making, so they started to innovate with palm oil, coconut oil, and eventually cottonseed oil, which was cheap and abundant. To ensure a constant supply of cottonseed oil, Procter & Gamble began to corner the market for it. This led to the company having a surplus of cottonseed oil.

In 1907, a German chemist sold to Procter & Gamble, the soap makers, the U.S. rights for a process that made a solid fat from a liquid (hydrogenation). The resulting colorless, odorless, tasteless fat made from the surplus cottonseed oil bore an uncanny resemblance to lard, so Procter & Gamble decided to sell it as a lard substitute. Crisco hit the market in 1910.

The genius of the marketers at Procter & Gamble was to introduce Crisco as a scientific breakthrough at a time when science was bringing all sorts of progress to consumers in the forms of electricity, transportation, and communications. To convince consumers of the value of Crisco, recipes and cookbooks were given away for free and the product was positioned as a "clean," healthy, more easily digested alternative to lard. Doughnuts were fried in Crisco and handed out on the streets. Samples were mailed to grocers, restaurateurs, nutritionists, and home economists. Never before were so much muscle and so many dollars put into a consumer campaign (and so the whole advertising industry was born). Consumers were hooked.

Sales rose from 2.6 million pounds of the stuff in 1912 to 60 million pounds in just four years. It wasn't until the 1990s that anyone began to question the health effects of the trans fats that make up 50 percent of Crisco and many other factory foods. According to *The Happiness Diet* by Drew Ramsey and Tyler Graham, "for every 2 percent increase in the consumption of trans fats (still found in many processed and fast foods) the risk of heart disease increases by 23 percent."

BIG FOOD INFLUENCES HEALTH POLICY

The history of how the lipid theory shaped heart disease research, and the research itself, is brilliantly documented by Nina Teicholz in *The Big Fat Surprise* (see Resources, page 278). This 481-page book takes a hard look at all the studies involved in the heart disease–diet connection and draws some conclusions that are quite different from what the American Heart Association (AHA) still preaches. This makes more sense when you understand how much "big food" has influenced American health policy.

That influence can be disguised as science. First came the Nutrition Foundation, created by Quaker Oats and the Corn Products Refining Corporation in 1941. This organization, writes Teicholz, "steered the course of science at its very source by developing relationships with academic researchers, funding important scientific conferences, and funneling millions of dollars into research." In 1948, Procter & Gamble took that strategy a step further: It donated all the profits from its popular *Truth or Consequences* radio program to the American Heart Association, then a small, underfunded organization founded in 1924 by cardiologists seeking to understand the growing problem of heart disease. With contributions from other food giants over the years, the AHA grew into the largest nonprofit in the country, with a $30 million budget by 1960. Is it any wonder that the organization focused on proving a connection between consuming saturated fats from animals and heart disease?

As early as 1964, there was a growing body of evidence to show that saturated fats were not the main culprit in rising heart disease; it was sugar. The sugar industry association worked hard to counter or suppress such claims. Among other things, they paid Harvard researchers for an article reviewing the scientific literature, supplying materials they wanted reviewed, and receiving drafts of the article to review before publication. The resulting article, published in 1967, concluded there was "no doubt" that reducing cholesterol and saturated fat was the only dietary intervention needed to prevent heart disease.

Beginning in the 1980s, finally, results from the first clinical trials on the health effects of low-fat diets started coming in. One study after another failed to show that lower-fat diets lead to lower levels of heart disease. And, although the vegetable oil and processed-food industries spent millions trying to counter the studies, more studies failed to prove diets lower in saturated fats reduced heart disease. This bears repeating: *There is no evidence to prove that diets lower in animal fats reduce the risk of heart disease.*

The American Heart Association started recommending substituting margarine, a trans fat, for butter in 1960 as part of "a prudent diet." It had no impact on heart disease rates, but so embedded was the anti-fat consensus that policy experts failed to reexamine their recommendations until the 1990s, when studies began to link the consumption of trans fats to heart disease.

In recent years, the vegetable oil companies have been experimenting with fats made using a process called interesterification, which, Teicholz reports, produces various fatty acids with unknown health effects, other than the fats don't seem to affect cholesterol levels. One small study has shown that interesterification of oils causes a 20 percent increase in blood sugars. But the focus on cholesterol continues on a policy level, even as the diabetes epidemic rages out of control.

And what oils are they interesterifying for the reformulated Crisco? As of 2012, Crisco consists of a blend of soybean oil, fully hydrogenated palm oil, and partially hydrogenated palm and soybean oils. Yum.

One of the leading researchers into the dangers of trans fats was Fred A. Kummerow, a professor of biochemistry at the University of Illinois at Urbana-Champaign. He was suspicious of trans fats, simply because they weren't natural. (There are some trans fats found in the milk, meat, and fat of sheep and cattle, but these are chemically different in structure.) Kummerow's first study, which found that trans fats accumulated in the liver, arteries, fat tissues, and heart, was published in 1957, 60 years before artificial trans fats were finally banned. How long will it take for the effects of interesterification to be revealed? Shouldn't we be suspect that these new fatty acids, not found in nature, may also be problematic?

At this point, studies finding no link between saturated fat consumption and heart disease should have convinced the USDA that 50 years of official policy has been wrong. There is no evidence to support cardiovascular guidelines that encourage high consumption of polyunsaturated fats and low consumption of saturated fats. Yet as recently as 2015, the dietary guidelines released by the USDA continued to advocate a low-fat diet, with most of the fat coming from monounsaturated and polyunsaturated fats. The USDA's MYPlate website states: "Healthy eating patterns limit saturated and trans fats. Less than 10% of your daily calories should come from saturated fats. Foods that are high in saturated fat include butter, whole milk, meats that are not labeled as lean, and tropical oils such as coconut and palm oil. Saturated fats should be replaced with unsaturated fats, such as canola or olive oil."

So despite the fact that the entire lipid theory of heart disease has been debunked, low-fat diets are still recommended and saturated fats are still stigmatized; indeed they are linked with artificial trans fats as though they offer the same negative health consequences.

There is no evidence to prove that diets lower in animal
fats reduce the risk of heart disease.

It's no wonder that people are afraid of lard and tallow, when they really should be questioning the health effects of polyunsaturated fats and interesterified oils. There is a large body of evidence that shows when vegetable seed oils high in polyunsaturated fat are heated, their unsaturated, unstable atomic bonds oxidize, creating toxic breakdown products, including aldehydes, formaldehyde, free radicals, and degraded triglycerides. How toxic? Toxic enough to kill rats fed diets of heated soybean oil in one study. One particular aldehyde is implicated in the development of neurodegenerative diseases such as Alzheimer's. Teicholz reports, "A pathologist from Columbia University also reported that rats fed 'mildly oxidized' oils suffered liver damage and heart lesions, compared to rats fed tallow, lard, dairy fats, and chicken fat, which showed no damage. Most of this research was published in obscure, highly technical journals that nutrition experts rarely read, however; and in the U.S., diet-and-disease researchers were instead focused almost exclusively on cholesterol, anyway."

WHAT ABOUT CHOLESTEROL?

Cholesterol comes from two sources. Your liver makes all the cholesterol you need; the rest comes from the animal products you eat. The less cholesterol you eat, the more your body produces. Cholesterol is transported through your bloodstream by carriers made of lipids (fats) and proteins — or lipoproteins.

From the 1950s through the 1980s, people were told to lower their total cholesterol levels, to eat the cholesterol equivalent of one egg a day. But such advice didn't lower the rates of heart disease. Then scientists turned their attention to the fact that there are two type of lipoproteins: low-density lipoprotein, or LDL, and high-density lipoprotein, or HDL. (There are also VLDL lipoproteins, or very low-density lipoproteins, but they don't play much of a role in what the public has been taught about heart disease.)

The common "wisdom" we have been taught is that LDL cholesterol is "bad" cholesterol because it contributes to plaque being deposited in arteries, which in turn raises the risk for heart attack and strokes. HDL cholesterol is "good" cholesterol because it carries LDL cholesterol away from the arteries and back to the liver, where it is broken down to be passed from the body. Consuming animal fats tends to raise LDL and HDL levels equally — making them a wash in terms of leading to heart disease.

The picture is even more complex than that. It turns out that not all LDL cholesterol is "bad." There are two types of LDL cholesterol: the large, fluffy type and the smaller, denser type. Eating saturated fat results in an increase in the large, harmless molecules, while eating carbohydrates increases the small, dense molecules of cholesterol — the kind that accumulates in the arteries. Unfortunately, there are no inexpensive tests to determine what size LDL molecules may be in a person's blood, so this fact is not widely disseminated to the public. But it does explain why some people have high cholesterol levels but suffer no signs of heart disease.

But wait, there's more. While many studies show that eating more polyunsaturated fats does lower cholesterol levels (but not affect rates of death by heart disease), it does seem to increase rates of cancer, particularly colon cancer.

Clinical trials have shown that polyunsaturated fats can negatively affect markers of inflammation, immune function, tumor biology, and prognosis.

Animal fats from pasture-raised animals begin to look better and better.

SOME FAT IN THE DIET IS NECESSARY

We need to eat some fat because our bodies cannot synthesize all the fatty acids needed for tissue health. The two primary essential fatty acids (ones we must consume) are linoleic acid (omega-6) and alpha-linoleic acid (omega-3). These fatty acids are necessary for the formation of healthy cell membranes, proper development and functioning of the brain and nervous system, proper thyroid and adrenal activity, hormone production, regulation of blood pressure, liver function, immune and inflammatory responses, regulation of blood clotting, transport and breakdown of cholesterol, and healthy skin and hair.

It is thought that humans evolved on a diet in which the ratio of omega-6 to omege-3 fatty acids was about 1 to 1, whereas in a typical Western diet that is high in processed foods, the ratio is about 10 to 1 or 15 to 1, depending on the population studied. What happens when the omega-6s and omega-3s are so out of whack?

Many of the chronic health conditions — cardiovascular disease, diabetes, cancer, obesity, autoimmune diseases, rheumatoid arthritis, asthma, and depression — are all associated with increased omega-6 fatty acid intake relative to omega-3 fatty acid intake. How to correct the balance? Decrease the amount of fat you consume that is high in omega-6 fatty acids, and increase the amount of fat you get from sources that are rich in omega-3s. Not surprisingly, this means enjoying more grass-fed, pasture-raised meat and eggs and fats and staying away from the polyunsaturated fats in seed oils. Industrially raised meats (and their fats) that come from grain-fed animals are not good sources of omega-3s. According to the Eatwild website (see Resources, page 278), which defends its statements with links to the original studies, "Meat from grass-fed animals has two to four times more omega-3 fatty acids than meat from grain-fed animals. . . . People who have ample amounts of omega-3s in their diet are less likely to have high blood pressure or an irregular heartbeat. Remarkably, they are 50 percent less likely to suffer a heart attack. Omega-3s are essential for your brain as well. People with a diet rich in omega-3s are less likely to suffer from depression, schizophrenia, attention deficit disorder (hyperactivity), or Alzheimer's disease."

Animal fats also transport and enable the use of fat-soluble important nutrients, including vitamins A, D, E, and K. The fats carry the vitamins through the intestine, into the bloodstream, and then to the liver, where they're stored until the body needs them. Vitamin A is essential for good vision, D for bone health, E for limiting the formation of harmful free radicals, and K for blood clotting.

Are you ready to start cooking and baking with animal fats like your ancestors did?

GRASS–FED ANIMALS ARE BETTER

There is plenty of truth in the old saying, "You are what you eat." For the animals we eat, this is also true. Cows do best on grass in terms of both their health and yours. When a cow grazes on pasture, its fat composition ends up with a ratio of two parts omega-6 fatty acids to one part omega-3, which is pretty much the ideal.

Every aspect of raising commodity cattle is about reducing costs, especially in terms of feed. This cattle is fed (probably genetically modified) corn and soy and will require antibiotics to reduce the incidence of acidosis, a painful, sometimes fatal affliction brought on by a diet that is too high in starch and too low in roughage. To further cut costs, the farmer may also feed the animals "by-product feedstuff," such as municipal garbage, stale pastry, chicken feathers, and candy. Studies on grain-fed steer found the ratio of omega-6 to omega-3 fats was between 5 to 1 and 13 to 1, which is far from the ideal.

Omega-3 fatty acids are the ones found in abundance in seafood (hence the current demand on our salmon stocks), nuts (hence expanded almond production in drought-stricken California, where it takes more than 1 gallon of water to produce one nut), and flaxseeds (whose nutritional benefits may or may not be fully available to humans). Omega-3s are also found in animals raised on pasture, which is not surprising because omega-3s are formed in the chloroplasts of green leaves and algae. According to the Eatwild website (see Resources, page 278): "Sixty percent of the fatty acids in grass are omega-3s. When cattle are taken off omega-3 rich grass and shipped to a feedlot to be fattened on omega-3 poor grain, they begin losing their store of this beneficial fat. Each day that an animal spends in the feedlot, its supply of omega-3s is diminished."

Another benefit of meat and fat from pasture-raised ruminants, such as cows, goats, and sheep, is conjugated linoleic acid (CLA). This is a naturally occurring trans fatty acid made from omega-6 essential fatty acids. Although not an essential nutrient we must consume to stay alive, CLA has been shown to have many health benefits, including protecting the heart; preventing thickening of arteries; preventing cancer, particularly breast cancer; and helping to regulate the immune system.

Among the many changes in our diet as we switched from a whole-foods, farm-based diet to a processed-foods diet, there has been about an 80 percent reduction in the amount of CLA we take in, at the same time that heart disease, autoimmune disorders, and cancer have been on the rise. Research by the University of Wisconsin shows that the best source of CLA is dairy and meat products from pastured animals, because they produce 300 to 500 percent more CLA than grain-fed animals.

Despite efforts to find ways to artificially raise CLA levels in animals, the only thing found to significantly raise CLA is fresh grass. CLA supplements are usually derived from chemically altered vegetable oils and have not been proven to be effective dietary supplements for animals or humans.

CHAPTER 2

Buying, Rendering, and Curing Animal Fats

The fat that is removed from the carcass of an animal must be rendered or cured before you can cook with it (with a few exceptions discussed below). Rendering is a process by which fat is melted, then strained to remove impurities and nonfat solids, greatly extending its shelf life. Curing is a process of preserving fat by just salting (salt pork), salting and smoking (bacon), or salting and drying (lardo, guanciale, pancetta).

This fat should, of course, be from pasture-raised animals because such fat is healthier for you and has different characteristics than fat from animals raised in concentrated animal feeding operations (CAFOs). Animals fed diets high in polyunsaturated fats — for example, hogs fed processed food wastes — will produce fats that are higher in polyunsaturated fat than pasture-raised animals. This fat will be softer, even when chilled. It will be especially unsuited for curing.

BUYING ANIMAL FATS

There are dozens of considerations when deciding which animals you might raise on your farm or homestead, including amount of land or pasturage available, time available, and ease or intensity of management, so chances are cooking-fat qualities won't be a primary concern. But if you are approaching animal fat as something to buy, then there are lots of factors that might affect your choices.

Be mindful that some animal fats you could be cooking with may be fats that you usually trim off and discard. I save up the fat from chicken parts that I have trimmed and store it in the freezer. When I have a pound or so, I render it. My son works as a line cook in a café where a lot of poultry fat is trimmed from birds; he saves it for me or shares it with friends. Perhaps you know someone who would save fat for you? It doesn't hurt to ask at restaurants and butcher shops. I'd like to think that in the future, no one will give away animal fats for free, but we aren't there yet!

You can also save fats from meats and poultry that you have cooked — the process of cooking creates rendered fat. (More on this later in the chapter.)

ONLINE OR LOCAL. You can buy already rendered fats from pasture-raised animals online (see Resources, page 278) and from some farmers' markets or butcher shops — much depends on the license the business holds and the state food-processing laws they are subject to. You may be able to buy unrendered lard and beef suet at butcher shops, especially if you call ahead, and directly from farmers who sell pasture-raised meat. The rendered lard I've bought at a farmers' market was an under-the-table cash transaction.

Duck fat and sometimes goose fat are available already rendered. Rendered duck fat is so commonplace that even Walmart carries it online. I've been able to find D'Artagnan brand rendered duck fat at my local coop and at other natural food stores. It may be possible to purchase unrendered poultry fats, but I have never found any to buy.

Beef suet, the hard fat around the kidneys, when rendered becomes beef tallow. Lamb and goat suet, also the hard fat around the kidneys, when rendered becomes tallow.

Lard is pork fat, rendered or not. Leaf lard is specifically the fat that encases the kidneys of pigs, rendered or not. When baking, leaf lard is preferred, because it has a higher melting point and often a less porky flavor than other pork fats. It is a firmer fat than lard from other parts of the pig.

Lard that is located along the back and upper sides of a pig is called fatback. It may have the tough skin (rind) on it, or it may not. Fatback can be cured to make salt pork. Some cooks prefer uncured, rendered fatback for frying because it does have a porky essence.

Bacon can be made from the pork belly and is sometimes called side bacon. Salt pork can also be made from belly fat. Pork jowls, a fatty cut, can be cured and smoked to make bacon. In Italy it is cured then dried but not smoked and is called guanciale. Guanciale is more porky in flavor than pancetta, which is a cured unsmoked bacon made from the belly.

Manteca is the Spanish word for lard, but it may also mean butter or peanut butter.

Chicken fat when rendered may be called schmaltz, which is what it is called in German and Yiddish. In fact, *schmaltz* in Yiddish means "poultry fat" but it encompasses all rendered animal fat in German (usually lard and goose fat). It is *graisse de poulet* in French, but it is not as common in the French kitchen as duck fat, or *graisse de canard*.

Goose fat when rendered is goose fat. It is *Gänseschmalz* or just *Schmalz* in German and *graisse d'oie* in French. Goose fat is more common than duck fat in German cuisine.

ALREADY RENDERED. The advantage of buying already rendered fats is that it is convenient, but it is much more expensive to buy it that way. If it is not more expensive, you may not be getting a high-quality product. At least that's true of lard; do not buy the shelf-stable pork lard that you can find in blocks or tubs unrefrigerated in the grocery store. Usually, this lard has been hydrogenated and/or otherwise altered to be stable without refrigeration. Scrutinize the label carefully and don't buy it if it contains anything other than 100 percent pure fat. You do not want any product that lists on its label "hydrogenated," or "BHA, propyl gallate, and citric acid added to help protect flavor." I don't recommend buying rendered lard that doesn't specify that it comes from pastured pigs. Also, if the lard doesn't specify "leaf" lard it may not be great for pastries, though it can be used for cooking.

Some brands of 100 percent rendered fat from pasture-raised animals that come in bottles or sealed plastic tubs say to "store in a cool, dry place." My personal opinion is that this fat should be refrigerated, just like the fat I render at home. Why take chances?

UNCOMMON FATS. Sometimes you can purchase rendered fats from less commonly raised animals. I have bought lamb tallow and bison tallow from online sources, and these have no culinary advantages over the beef suet that I can easily source locally. I have also bought rendered beef tallow, lard, leaf lard, chicken fat, duck fat, and goose fat online and, again, these have offered no advantages over the fats I have rendered myself. These online products, however, are a good way to try out different fats without investing too much time or effort.

CURED FATS. Salt pork and bacon are two commonly cured fats; they are easily found in supermarkets, though the quality will vary tremendously. Purchasing commodity bacon and salt pork is much cheaper than making your own, but if you are working with pasture-raised pork, the difference in quality is obvious. Italian-style cured fats, such as pancetta and guanciale, are easily found online or at specialty food stores and are usually pretty expensive. In the South, it is possible to find cured side meat, which is cured but not smoked bacon (pork bellies are also called sides).

CAUL FAT. This is the only fat I know of that is easily cooked without rendering or curing. Caul fat (also called lace fat, mesentery, or crepine) is a thin, lacy, fatty membrane from the lining of the stomach of beef, pigs, sheep, and other ruminants, such as venison. It is mostly used by chefs to wrap sausages, roasts, roulades, crepinettes, and lean game birds. The caul holds these items together while melting into the surface of the food, keeping it moist. Pork caul fat adds a slight bacony flavor, while caul fat from sheep and cows is more neutral in flavor. It is possible to find caul fat online, in bulk quantities, but it must be shipped frozen because it is highly perishable. I haven't found a local source and so have little experience with it.

UNRENDERED FAT. You can find some recipes that say you can use grated, but not rendered, beef suet, as long as you remove all the impurities. Good to know, but there are a lot of impurities to be removed from a typical hunk of beef suet. I prefer to render it first.

Unrendered backfat from Mangalitsa (a famously fatty breed of pig from Hungary) pigs is sometimes finely ground and then whipped into a spread that is enjoyed on bread (see Whipped Lardo, page 104). The lard from Mangalitsa pigs is sweeter and lighter and melts at a lower temperature than lard from other pig breeds, because it contains more unsaturated fat.

Beef suet is shot through with collagen. It is easily broken apart but must still be chopped to make pieces small enough to melt quickly.

ANIMAL FAT CHARACTERISTICS

If you're raising beef cattle, you will have an abundance of beef suet; if you raise lamb, you can reserve some suet for cooking with; if you raise chickens . . . well, you get the point. But assuming you have to buy the fat, how do you choose what to buy?

Flavor, Texture, and Mouthfeel

In the most general terms, poultry fats and cured pork fats add the most flavor. All rendered fats add great texture and a silken mouthfeel. Adding poultry fat in place of butter to mashed potatoes, winter squash purées, or steamed or sautéed vegetables adds a barely discernible hint of poultry, but more than that, it adds a luxurious quality. Sautéed kale in poultry fat is simply delicious. Just as your morning eggs have more flavor when fried in bacon fat, they will have more flavor when cooked in poultry fats. Adding salt pork or bacon fat to bean dishes, soups, and braises adds great flavor; using uncured fats won't make much difference.

When starting a recipe by sautéing some aromatics (garlic, onion, celery, parsley), any animal fat works. Any animal fat will work in a stir-fry, though lard is most traditional. When making a roux to thicken a sauce or make a gumbo, any animal fat will work, though it is pleasing to use chicken fat for chicken dishes, tallow for beef dishes, and so on.

WHAT COLOR IS YOUR FAT?

Tallow and poultry fats appear golden when melted; lard is a translucent, milky white. When tallow, lard, goose fat, and duck fat become solid, they turn white. If your solid fat — other than chicken — has a yellowish cast, it reflects the diet of the animal, which was probably fed a lot of corn and other grain (not grass). Chicken fat never becomes firmly solid and remains yellow in color. If your fat — of any origin — turns beige or brown, chances are it has been overheated, and it should be discarded.

Flaky Texture for Baked Goods

Lard and tallow add flaky tenderness to pastries and allow cookies to bake up higher due to the fats' higher melting temperatures relative to butter. Lard has an edge over tallow when it comes to making dough that is easy to handle. Doughs can also be made with poultry fats. They will not yield quite as flaky a product, but they will make far tastier, flakier, crispier doughs than could be achieved with vegetable seed oils, and be far easier to handle. Piecrusts made with animal fats seem to better resist becoming soggy over time compared to butter-based piecrusts.

Lard makes the crust of this Beef and Mushroom Pie (page 177) especially flaky.

Lighter Frying

As a frying medium, animal fats produce far less greasy results than vegetable oils whenever and whatever you are panfrying or deep-frying. Frying potato pancakes, crab cakes, and vegetable fritters in animal fats is a revelation because the cakes turn out crispy and not at all greasy. The flavor beef tallow and lard add is subtle, but good. When McDonald's, bowing to pressure, started frying their fries in hydrogenated vegetable oils rather than tallow in 1990, they started adding "natural beef flavor" to make up for the loss of flavor. Poultry fats are also wonderful, nongreasy mediums for frying — though it is difficult to acquire them in the quantities required for deep-frying. (Ryan Rogers of Royal's Hot Chicken fame in Louisville, Kentucky, started out with the idea of frying chicken in chicken fat. But after his first month in business, there was no more chicken fat to be had within a 100-mile radius.)

WHICH FAT IS FOR YOU?

More cuisines have been almost entirely dependent on lard as their go-to fat than any other fat, undoubtedly because pigs are easier to raise and require less grazing land than cattle, and they are so generously endowed with fat. It is also a go-to fat for many because it is excellent for both stovetop cooking and oven baking.

Although sheep tallow can be used wherever beef tallow is used, the leaf suet on lambs raised for meat often goes into the ground meat and sausages, with none left over (there's only about a pound of leaf suet to work with). It's the same thing with goats. Since goat meat is so notoriously lean, much of the leaf suet goes into ground goat. However, if you raise goats for meat, you can ask your butcher to reserve the fat in the body cavity for cooking purposes.

Some people complain that beef tallow is too waxy, or that it leaves a waxy coating on the palate. I will note that the fat around the heart of all animals tends to be waxy, but I have never encountered the waxy quality that others talk about. Indeed, when I have made chocolate chip cookies from tallow and from lard, tasters couldn't tell them apart.

Search for "bear fat" online and you will inevitably come across Michelin-starred chef April Bloomfield's recipe for potatoes crisped in black bear fat, complete with photos. If that isn't proof that bear fat can be cooked with delicious results, I don't know what proof you would require.

There is plenty of folklore surrounding bear fat and plenty of recorded uses, including lubricating guns and conditioning and waterproofing leather and wood. It has been used as lamp oil and as an insect repellant. Rubbed on the scalp, it is said to repair or prevent hair loss. I'll take that one with a grain of salt, but I'm pretty sure that it is a great cooking medium when properly handled and rendered. And that's where you might expect problems to come into play. Any time you hear that bear grease is rancid or taints the meat, you can be pretty sure that the bear fat has been handled improperly and probably is rancid.

Black bears are hunted in fall. The later in the fall, the thicker the insulating fat layer — from 2 to 5 inches thick. To prevent the meat and fat from becoming rancid, the bear should be field dressed and skinned immediately. Then the fat — that "insulating layer" — should be removed by slicing it off in strips. If the weather is cool, the fat will be solid; if the weather is warm, the fat will be like a jelly. In either case, the fat should be rapidly chilled and kept as cold as possible.

To render the fat, treat it just like any other fat. Trim away any flesh (or hair). Cut it into the smallest pieces possible and render it in a heavy pot with a small amount of water to prevent scorching. As with beef tallow, all you want to do is melt the fat to separate the fat from the solids. Don't let the fat sit on the solids as they color. Strain the fat to remove any solids and store in the refrigerator or freezer.

And don't forget that bear fat, like the fat of all animals, is affected by diet. A bear that has been feeding on berries, field crops, and nuts will taste different (and far better) than an animal that has been feasting on garbage at the town dump.

So which animal fat should you use? For many of the recipes in this book, I have called for "any animal fat" where any would work. Generally, all the poultry fats are used interchangeably, though goose fat and duck fat offer a somewhat richer or more enhanced flavor than chicken fat. Lard and tallow are virtually interchangeable when it comes to deep-frying and stovetop cooking, though not for baking.

I try to keep either rendered lard or tallow in the house, depending on what I have been able to source and when. And I try to keep some rendered poultry fat on hand. Since I learned how to cook ducks and geese properly, I buy one or the other when I can. Chicken is available year-round, as are ducks; geese are not commonly raised in this country and tend to be available only from late fall to early winter (they are harvested and sold frozen for Christmas). I also save chicken fat trimmed from the dark meat pieces we prefer and render it regularly.

ACQUIRING ANIMAL FATS TO RENDER OR CURE

If you raise the animals yourself and slaughter and butcher them yourself, you can make sure all the leaf fat or cavity fat is retained for you to render (freeze it until you are ready to work with it), or you can make sure the slaughterhouse or butcher packages the fat for you. Unless you specifically ask your butcher for the kidney fat of ruminants (the fat around the internal organs), it will be used in sausage making, added to ground meat mixtures, used to make animal feed, or discarded. So make sure your butcher knows how much you value "leaf lard" in a pig and "leaf suet" from cattle, sheep, or goats. The fat from muscle trim has different cooking properties — it is softer — and is less commonly saved for rendering, though it can be; it can also be cured.

If you don't raise animals yourself, you will need to find a local source of pasture-raised animals — either from a butcher or directly from a farmer. When you buy a side of beef or a half pig or a whole lamb, you pay for the hanging weight, after the skin and guts are removed. There should not be an extra charge for any fat you ask to be reserved; you have already paid for it, so why waste it?

Beef, Lamb, and Goat Fat

The fat from beef, sheep, and other ruminants is called suet in its unrendered state and tallow in its rendered state. Tallow from different sources has very similar characteristics for cooking. The difference is quantity. There is so much more from a steer — anywhere from 10 to 20 pounds of useful leaf suet, depending on the breed, age, and diet of the animal. (There is also subcutaneous fat, just under the skin, and fat between and within the muscles. This "soft fat" is left on the cuts that you cook.)

A whole lamb will yield about 1 pound of leaf suet, and a mature ewe will yield more, but it is difficult to acquire. There isn't much of a market (or any market) for mutton (mature sheep) in this country, so ewes that are slaughtered because they are too old to continue breeding are generally turned into sausage, with all the fat going into the sausage formula. The leaf suet from lambs and goats is generally added to ground meat mixtures, with none left over, so this discussion is mainly about beef fat.

Some of the fat will be ground with tough cuts and trim to make ground meat and sausage, but plenty should still be left. Ask for the suet to be packaged in 1- or 2-pound packages. Otherwise you may end up with an unwieldy 15-pound package.

Beef suet looks like weird stuff. It is an irregularly shaped block of fat — all uneven surfaces and odd angles held together with thin, transparent ribbons of collagen. Depending on the steer's diet, the fat will be mostly pure white or it may be yellowed. It will have some pink areas and may have some congealed blood, which you should wipe away. It will have the same waxy texture whether frozen or chilled and should have no odor.

Rendered suet becomes tallow and is white and quite brittle when cooled. It has a high melting point, making it excellent for pastries, and a high smoke point, which makes it very good for deep-frying. If rendered properly, the rendered fat should be entirely free of beefy or muttony flavor.

Store suet in the freezer until you are ready to render it; it keeps, unrendered, in the fridge for only 5 days. Because it has the same texture frozen or unfrozen, there is no need to thaw it before rendering. I generally get 1½ to

1¾ cups of tallow for every pound of suet I render. The impurities in tallow (the cracklings) aren't particularly tasty, so they can be fed to animals or discarded (my neighbor's chickens enjoy them). Rendered and cooled tallow cannot be easily spooned from a jar. It will keep for about 6 months in the refrigerator and much longer in the freezer.

FAT MATH

When you order rendered fat online or buy it in a butcher shop or specialty food store, you pay a premium price (for fat from pasture-raised animals). But when you render it yourself, how much usable fat do you actually get? It comes out very, very roughly as ½ pound unrendered fat yields 1 cup rendered fat or 7 ounces. When I render beef suet, I do not extract every last bit of fat; rather, I leave some unrendered once the cracklings start to brown because I want to keep it very neutral in flavor. Here are some approximate measures, based on my experience:

ANIMAL	AMOUNT FAT	AMOUNT RENDERED FAT	AMOUNT CRACKLINGS*
9-pound goose	2¾ pounds	4¾ cups	1 cup
7-pound duck	8 ounces	1 cup	½ cup
3 to 4 sides of leaf lard	6¾ pounds	12 cups	1 quart
Beef suet	4½ pounds	5 to 7 cups	None

* The impurities that are rendered out from pigs and poultry are called cracklings, and they are generally quite tasty when crisped in a dry skillet and salted. They are good for adding to corn bread and for topping salads. Cracklings from rendering beef suet do not taste good but can be mixed into animal food.

RENDERED BEEF TALLOW

UNRENDERED BEEF SUET

WHAT ABOUT DEER FAT?

"What about deer fat — can I cook with it?" This is a question I hear frequently, and the answer is a qualified yes. Deer fat, like that of all ruminants, is suet in its unrendered state and tallow when rendered. Venison is famously lean, but there is fat found around the internal organs that is good to cook with, provided that fat is very rapidly chilled after the deer is slaughtered in the field — and therein lies the rub.

A deer taken in the wild is quickly eviscerated, and guts are usually discarded right then and there. If you want to save the deer suet, you need to cut away the fat from around the kidneys. This is the leaf suet and the fat most valuable to cook with. It needs to be covered and chilled very rapidly.

The body fat is often exposed to air as the carcass is cooled, which causes the fat to oxidize (become rancid). This is why hunters are urged to get rid of all the fat and told that it doesn't taste good. Rancid fat never does.

You may be told that venison fat has a high melting temperature and will leave a waxy coating in your mouth. In fact, venison fat has the same melting temperature as beef (112° to 115°F) so it is unlikely you would perceive any waxy coating — nobody complained about a waxy coating left by McDonald's French fries for all those years they used tallow in their fryers.

Pork Fat

A pig carcass contains leaf lard around the kidneys, jowl fat, fatback (or backfat), belly fat, and intermuscular and intramuscular fat. The leaf lard is a hard fat and is most prized for making pastries. Jowl fat, fatback, and belly fat are soft fats that can be cured to make guanciale, salt pork, bacon, and other cured products. The fat on and around muscles is either ground up and used in ground pork and sausage or it is left with the meat cuts to moisten the meat as it cooks. With a 250-pound pig, depending on the breed, there will be 4 to 8 pounds of leaf lard and another 4 to 8 pounds of fatback and pork belly.

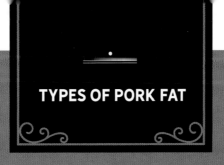

TYPES OF PORK FAT

All pork fat is not created equal. When asking for fat from a butcher, or filling out a cut sheet for a slaughterhouse, be sure to ask for the types of fats to be packaged separately because you will be using them differently.

FATBACK (also called backfat). This is the fat that comes from the back of the pig along its spine from shoulder to rump. It's literally the layer of fat directly below the skin of the back. It's usually sold in pieces and may or may not have the skin still attached. If you want the skin, be sure to specifically ask for it, otherwise it may be trimmed off. Fatback can be cured to make salt pork, with or without the skin attached. Rendered fatback is great for sautéing and frying (but has too much pork flavor for pastry). It is considered a soft fat.

BELLY. This is the flesh that runs on the underside (the belly) of the pig and surrounds the stomach. It is one long cut of meat with plenty of fat layered with the meat, which is why it is prized for curing into bacon, pancetta, or salt pork. The belly can also be roasted or braised and then broiled for a crisp finish.

LEAF LARD. Leaf lard is the fat from around the pig's kidneys. It is mostly pure fat, with some collagen shot throughout. This is the fat that you want to render into a pure white fat to bake with. Leaf lard is a hard fat.

LEAF LARD

PORK BELLY

RENDERED LARD

FATBACK

Leaf lard comes in one long length. It will be white with tinges of pink. Like suet, it contains thin, semitransparent ribbons of collagen, and there may be bits of blood, which should be wiped off. It should have no odor and should be stored in the freezer until you are ready to render it. Unrendered leaf lard keeps in the fridge for only 5 days. It is easier to handle when frozen; when thawed, it will leave a greasy film on everything it touches.

I generally get 1½ to 1¾ cups of lard for every pound of leaf lard I render (plus cracklings). I recently rendered about 8 pounds to make 10 cups of pure white leaf lard (good for baking), 1½ cups rather porky lard (good for stovetop cooking), and 3 cups cracklings.

Melted lard is a translucent white to slightly golden color; it is pure white when it solidifies. At room temperature, it has a creamy texture, very much like factory-made vegetable shortening, and is fairly easy to spoon out of a jar. It is harder to spoon out of a jar when chilled. It may have a faint porky, but not offensive, odor. If it smells very porky or is not pure white when solid, then it sat on the cracklings for too long as it rendered.

WHY NOT USE SUPERMARKET LARD?

The shelf-stable lard that you find in the baking aisle of the supermarket is not recommended. Armour, which is the most readily available brand, contains "lard and fully hydrogenated lard, BHA, propyl gallate, and citric acid." Armour's website includes this information: "Armour lard is fully hydrogenated and contains zero trans fats. Fully hydrogenated lard is different than partially hydrogenated oils, which are chemically altered and contain trans fatty acids. Trans fats are harmful because they lower HDL (good) cholesterol and raise LDL (bad) cholesterol levels, increasing the risk of coronary heart disease." A fully hydrogenated fat is 100 percent saturated fat, whereas lard normally contains 45 percent monounsaturated fat and 11 percent polyunsaturated fat — and no preservatives.

Poultry Fats

The most easily removed fat from poultry is generally found in the cavity near the neck and bottom — and it is generally left in the bird by the farmer or slaughterhouse. Just reach in and pull out the blobs of yellow-white fat you find there. If you like, you can collect that fat in the freezer and add to it when you are trimming parts or skinning a bird. Save the skin to make cracklings when you render the fat off. These bits of skin are sometimes called gribenes or greaves, and they are delicious.

The amount of fat you can get from a bird varies tremendously depending on whether the farmer or processor has a side business in selling fat. It also varies

When collecting duck fat to render (left), save the excess skin, as well. Chickens may not yield as much fat (right). Rendered chicken fat is quite yellow and only semisolid.

depending on whether you break it down (cut it into parts) or leave it whole. A whole chicken weighing 5 to 7 pounds may yield less than an ounce of fat that you can pull from its cavity. Break down that same bird and trim the pieces and you can get several ounces. Before breaking down a 6-pound duck, I pulled 7.5 ounces of fat from it. After I broke down the bird, I had 8.3 ounces of fat to render. I have been able to render about 9 ounces of fat from a 7-pound goose.

Because pasture-raised chickens are often harvested at an older age than industrial-raised chickens, they may yield greater amounts of fat. That said, birds that aren't given a lot of supplemental grain can be quite lean. A breed like the Red Ranger is much leaner than a breed like a Cornish White, which has been bred to be a weight-gaining machine. With ducks and geese, I have found less fat on pasture-raised birds than on supermarket birds — this may be a function of breed or diet or both.

I save up the fat from every chicken or package of dark meat chicken I buy; I also save skins if I need skinless parts. I keep a large plastic bag in the freezer, and when I've accumulated a few pounds of fat, I render it. Ten chicken thighs, for example, will yield about 6 ounces of trimmed fat.

Every once in a while, I buy a duck or a goose. Both are tricky to cook if you don't want tough, stringy meat that is raw close to the bone. They are also quite expensive. Generally speaking, a whole duck yields one dinner for three plus about ½ cup of rendered fat. A fresh goose of 6 to 8 pounds is very expensive, and available only in the late fall through early winter, but may yield up to 2 quarts of fat (or liquid gold). (Wild geese are much leaner than farmed geese.)

Unrendered poultry fat is generally a yellowish white in color and has a soft, greasy texture. Like all animal fats, it is easiest to chop when it is partly frozen. Once rendered, it becomes a semisolid, easily spooned from a jar directly from the refrigerator. At room temperature, it is liquid.

You can often find rendered duck fat or goose fat at specialty food stores, often imported from France. Since no one has figured out how to raise ducks and geese in confinement, you do not have to be quite as careful with its origins as you should be with pork and beef fat.

SAVING FATS FROM COOKED DISHES

There's another source of fat that should not be overlooked. Every time you cook meat, you can retain the fat that is rendered off. This is the fat that forms a hardened shell on top of the soup or stew. It is also the fat that collects on top of the juices from a roast if you aren't making gravy. Just skim it off and store it; in the refrigerator it will keep for about 5 days — or as long as the original dish stays good. Put it in the freezer (labeled) for longer storage.

When you roast bones to make a brown stock, or when you roast them for the marrow, some fat will collect in the roasting pan, and this fat can be used for cooking. When you make confit with poultry (cook it in its own fat), the fat can be recycled for other (delicious!) dishes.

When making duck confit, you will have both juices (bottom layer) and fat (top layer). Once the two layers have solidified, the fat layer can be peeled off the layer of gelled juices.

All of this collected fat retains the flavor of the original dish, and it is just as perishable as the dish it is collected from. Use this fat for any stovetop cooking you might do — sautéing meats or vegetables, stir-fries, making a roux — where the flavor is appropriate; it won't be good for pastries. It makes a great cooking medium for any sautéed vegetable or rice pilaf–type dish, adding flavor and depth. When I make gravies from any roast, I always make a roux from the fat that has rendered out of the meat.

To extend the life of such fat, you need to separate the fat from the juices that cling to it. A simple way to do this is to pour the fat and juices into a shallow plastic container or bowl and refrigerate overnight. The fat will form a solid cap on the top. Spoon off the fat, place the fat and juices in separate containers, and freeze or use within five days. Alternatively, pour the fat and juices into a shallow container with a tight-fitting lid. Place it upside-down in the refrigerator overnight. The next day, invert the container onto a flat pan or dish (as shown in the photos on the right), so the fat is once again on top, then peel or scrape the fat off.

When you cook bacon, you are looking for crispy meat with all the fat rendered out. All that fat can be strained and saved to cook with (see page 60).

In the United States, we call the fat that drips from meat as it cooks drippings, and we use this fat as the base of a gravy. If we are particularly frugal — or fat savvy — we save these drippings for use instead of rendered tallow or lard in a stovetop dish. In the United Kingdom, "dripping" — singular — is a well-known and popular substance with tradition behind it. People speak fondly of toast and dripping (see "Drippings on Toast," page 109), particularly those who remember the austerity days of World War II and its aftermath. Dripping is the basis of Yorkshire pudding and was once a popular medium for frying fish. In the UK, one can buy "deodorized dripping" or full-flavored dripping.

A full-flavored dripping from James Whelan Butchers in Tipperary was named the Great Taste Supreme Champion 2015 out of 10,000 entries. One of the judges commented that with one bite (presumably spread on bread), one got the flavors of roast beef, and the fat on the beef, plus the potatoes, Yorkshire pudding, and gravy.

On receiving the award, Pat Whelan said, "Our team is very honored to receive the Golden Fork Award and overjoyed to win the highest accolade from the Guild of Fine Food, the Supreme Champion Trophy. Bringing beef dripping back into the market wouldn't have been as straightforward twenty or even fifteen years ago. Now over the last few years

with the diet police's crosshairs firmly on refined sugar and carbohydrates instead of saturated fat, our customers are educated and appreciate this classic flavor enhancer. Fat is an essential part of our diet, protecting organs and aiding the absorption of vitamins, and according to many scientists, we are probably better consuming animal fats like butter and dripping. It's an ideal time to reassess some of the old-fashioned foods many have dismissed."

HOW TO RENDER FATS

The only real drawback to rendering your own fats is time. Regardless of the type of fat you are working with, the process is pretty much the same. Chop the fat into small pieces (1-inch dice or smaller), put in a heavy pot, and add a small amount of water to prevent scorching until the fat begins to melt. Melt the fat over low to medium heat, stirring frequently; do not let it bubble furiously. When the solid bits begin to color, begin removing and straining the fat through a fine-mesh strainer into storage containers. Ideally, the fat will be entirely neutral in taste, but the longer the melted fat remains with the solid golden or browned bits, the meatier the fat will taste. If your fat tastes too "porky" or "beefy" to you, it has probably been allowed to sit with the browned bits for too long. So be vigilant and watch for browning.

Some people like to render in a covered roasting pan in a 250° to 300°F oven. In this way, they avoid the necessity of keeping watch over the rendering. It takes longer in the oven, and more odors accumulate in the house. It is also hard to judge when it is time to start pouring off the rendered fat, so there is the danger of letting the fat develop the meaty flavors I try to avoid.

Beefy, porky, or poultry odors do accumulate in the house when rendering fat. If you don't have good ventilation in the kitchen, and the odors are a

"DRY" VS. "WET" RENDERING

I recommend that you add ¼ to ⅓ cup of water to the pan when rendering chopped-up suet and lard to prevent the fat from scorching or sticking to the bottom of the pan. (With its low melting point, poultry fat doesn't need this.) As the fat renders, this water will evaporate. This is not "wet" rendering. Wet rendering is a commercial process that takes place in giant steam kettles and does not resemble the home process at all. So don't worry about that little bit of water.

problem for you, you can render the fat in a slow cooker set outside on a porch or deck (I do this sometimes). I start it at high heat, then reduce the heat to low once the fat starts to render. I leave the lid off so condensation doesn't drip back into the fat.

I find that rendering on top of the stove is quicker than in the oven, resulting in less odor. The only drawback to rendering on the stovetop is that there will occasionally be a miniature steam explosion that will "pop," causing a small spatter of grease that can burn you if you happen to be standing near.

Batch sizes don't matter. Render a little at a time or render in big batches — whatever works for you. However, if you are working in big batches, ladle off fat as it renders to prevent flavors from developing.

Equipment

Whether you're rendering on top of the stove or in the oven, heavy cookware is recommended. Avoid lightweight pots, which can allow the fat to scorch. A slow cooker works fine.

KNIVES OR CLEAVERS

To get the purest, most neutral-tasting tallow or lard, you want to start with the smallest pieces possible; this enables you to quickly melt the fat and get it off the solid bits before they start browning and flavoring the fat. So before you start, consider first how to get your fat into small, quickly melted pieces. You can use a knife or cleaver on a cutting board. I use a cleaver for my chopping — and it takes me about an hour to chop up 8 pounds of suet, because it is so hard; chopping lard goes faster. I line my counters with old newspapers to make cleanup easier. When I am chopping, pieces of fat tend to fly around.

FOOD PROCESSOR OR MEAT GRINDER

Some people use a food processor with a grating disc and some use a meat grinder that's either freestanding or attached to a stand mixer. The food processor or meat grinder should be very well chilled before using; even then the works

will get gummed up — and the more moving parts, the greater the cleanup time. However, the fat will melt much faster if it's ground or grated, thus reducing the chance of picking up any meaty flavors. I've found that a powerful, freestanding meat grinder works best. The grinder attachment on my stand mixer tends to get gummed up, and the tough collagen plugs up the works. You'll still need a cutting board and cleaver to get the suet or lard into pieces small enough to fit into the chute.

LADLE, SPOON, AND STRAINER

You will need some sort of ladle to transfer the melted fat out of the pot and into a strainer to remove the remaining solids. You will also need a spoon for occasionally stirring the melting fat and something to rest the spoon on. If your strainer isn't a fine mesh, you can line it with paper coffee filters, but you'll go through a lot of filters as they become saturated. I don't recommend straining through cheesecloth or butter muslin for the simple reason that the cloths will have to be handwashed in extremely hot water and the wash water dumped outside; you can't wash fat-coated cloth in your washing machine unless you want to ruin your plumbing.

STORAGE CONTAINERS

Consider what sort of container will hold the strained fat — and how easy it will be both to store the container(s) and to extract the fat as you need it. Because tallow is so hard, I pour the melted fat into a glass 9- by 13-inch baking dish. Once the fat is solid, I can slice it, then remove it in strips with a spatula and store it in a flat plastic container. Lard and poultry fats are spoonable, so I put them into plastic deli containers or widemouthed canning jars.

If you are using glass canning jars, it is a good idea to keep them warm in a 200°F oven, as you would if you were canning jams or pickles. My canning jars are old and some have been used in the freezer; therefore they are more prone to breakage than brand-new jars. Once I had a canning jar break after filling it with hot goose fat. I don't know what was more painful — losing 2 cups of goose fat or the cleanup that followed. So now I hold my clean jars in a warm oven.

Cleanup Strategies

Finally, consider how you will clean everything, because you do not want the fat or fat residues to go down the kitchen sink and through your plumbing. Fat that turns solid at room temperature will solidify in the pipes, causing all sorts of problems. Try to keep dripping to a minimum to avoid having to clean up many different surfaces.

To clean the pot and utensils I've used, I heat a large pot of water to a boil. I dip all the utensils in the hot water, rub them clean with paper towels, then wash them. The pot in which I rendered the fat gets the remaining hot water sloshed in, then dumped out outside. Again, I rub it clean with paper towels, then wash it. I do use a lot of paper towels when I render fat. Otherwise, I take great pride in not using a lot of paper towels in my kitchen.

⩫ RENDERING ANIMAL FAT ⩫
AT A GLANCE

1 Chop the fat, if necessary. When you're working with tallow or lard, it's helpful if the fat is very cold — even frozen. To get even smaller pieces, use a well-chilled meat grinder or the grating disc of a food processor.

2 Transfer the chopped or ground fat to a heavy pot, adding just enough water to cover the bottom of the pot to prevent scorching.

3 Use low heat, ladling off the fat as it melts and straining it through fine mesh. Do not let the rendered fat sit on the browning bits, where it will pick up meaty flavors.

4 Discard the cracklings from tallow or feed them to pets. The cracklings from chicken and pork are delicious and can be crisped in a skillet, seasoned with salt, and eaten.

RENDERING BEEF SUET

Beef suet will have the same waxy texture whether frozen or chilled; at room temperature, it will have the same hardness but will leave a greasy film on everything it touches — your hands, the cleaver you chop it with, a meat grinder. It's best, therefore, to start with frozen suet and to work with chilled utensils to reduce it to small pieces.

A box grater works for a while, but the warmth of your hands will make the suet greasy, and it will eventually clog up the holes on the grater. I find the warmth of the motor of a food processor also starts to heat the suet and gum up the blade, but it works well, with many stops to clean up the works. If I've forgotten to chill my equipment, I stick with my cleaver, though it does make flying bits of suet inevitable and it does take some effort to chop it all up.

Work with a couple of pounds at a time. If the fat is in a big hunk, whack it a few times with a cleaver to get small enough pieces to work with. When you have enough small pieces to fill your heavy pot, start rendering. Add more chopped pieces as the volume decreases or wrap up any remaining suet and return it to the freezer.

Pour ¼ to ⅓ cup of water (enough to cover the bottom) into the pot at the start of rendering. This is to prevent scorching; it will evaporate before you have finished rendering. Keep an eye on the pot and stir occasionally.

Because I do not want beefy-tasting tallow, I start to ladle off the fat as it melts. It goes through a fine-mesh strainer and into a flat baking dish.

Rendered tallow is golden when hot, but it turns white when cooled and becomes quite solid at room temperature and brittle when chilled. It has a slight but distinctive odor, but should not smell particularly beefy. I want my tallow as neutral tasting as possible, so I don't extract every last drop and I discard the cracklings, though the cracklings make a fine addition to animal food, if you prefer.

Once hard, the tallow is easy to score with a knife and break into pieces, which I store in a plastic container. When I want to cook with it, I break off a piece. When I want to bake with it, I weigh it.

Pour the hot rendered beef, sheep, or goat fat into a flat baking dish and allow it to harden.

The hardened fat can be sliced into pieces and stored in plastic bags.

RENDERING LARD

As with beef, the lard should be chopped into small pieces and placed in a heavy pot with enough water to just cover the bottom of the pot to prevent scorching, then melted over low heat. I like to ladle off the fat as it melts and strain it into a jar.

At the end of the process, you will be left with some fairly wet and greasy brown bits. I transfer these to a skillet and continue to render and brown these cracklings until they are well browned. I strain these, to get every last ounce of lard, which will be beige in color and which I label "porky lard." This final ½ cup to 1 cup of lard is fine to cook with, but can't be used for pastries. I season the cracklings with salt and use them as a garnish or a snack.

Melted lard is translucent white at the beginning of the process and will take on a light gold cast as it spends time on the cracklings. I like to start ladling off the lard while it is still translucent white and use this, the purest lard, for pastry. The lard turns white when solid and creamy in texture. You should get about 1½ cups or 12 ounces of rendered lard plus about ½ cup cracklings per pound of unrendered lard.

Although all lard can be rendered, I render just the leaf lard and reserve the fatback and trim for salt pork. If you do render fatback or other lard, be sure to label it so you know to reserve it for cooking, not baking.

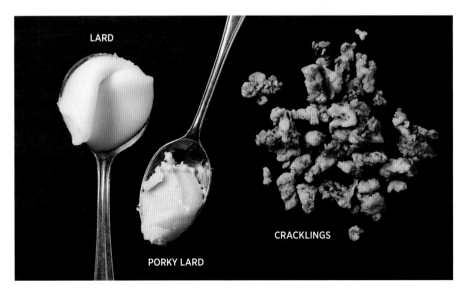

LARD

PORKY LARD

CRACKLINGS

Most of your rendered lard should be pure white and odorless. The lard you strain off the cracklings will be beige and have a distinctly "porky" odor.

CRACKLINGS

When you render any animal fat, you are left with cracklings. Cracklings are also what you get when you fry or roast pork skins. Either way, salted cracklings are delicious morsels of animal goodness. But a word of caution: They hold their heat, so don't snack on them hot out of the pan.

What can you do with salted cracklings? Corn bread with cracklings (see page 218) is an obvious choice. Many people add cracklings to salads as a crunchy topping, or incorporate them into egg dishes. The Italians call these crunchy bits *ciccioli* and serve them with wine and crackers as an aperitivo. In Poland, they are called *skwarki* and are served as a topping for pierogi and sauerkraut.

They are also incorporated in a spread made with lard called *smalec*. In Croatia and Serbia, pork cracklings are known as *čvarci*, and in Bulgaria, they are *prŭzhki*. In Serbia, they are incorporated into biscuits that are quite similar to our American biscuits.

In France, pork cracklings are *grattons*, and poultry cracklings are *fritons*. The French incorporate *grattons* into fougasse, which is a type of bread. In Yiddish, chicken cracklings are known as *gribenes* and often cooked down with onions and incorporated into potato dishes. At the Jewish delicatessen–inspired N4 taproom in Brooklyn, gribenes substitute for the bacon in a sandwich with lettuce and tomatoes, making it a kosher GLT.

RENDERING POULTRY FAT

Turkeys don't yield enough fat to allow for rendering, but chicken, ducks, and geese provide plenty, relative to their sizes. You will get about 1 cup of rendered poultry fat for every 1 pound of fat, plus about ¼ cup cracklings.

In Jewish cuisine the cracklings are cooked with diced onion until the onion bits are deeply browned. The resulting bits are salted and called gribenes and can be used for garnishes and snacking. They are ridiculously, dangerously delicious.

Like suet and lard, poultry fat should be kept frozen until you are ready to render it, and it should be cut into small pieces to render. It does not need the addition of a little water because it melts quickly.

Once rendered, poultry fat will keep for about 4 months in the refrigerator in a tightly sealed container or indefinitely in the freezer.

Rendered chicken fat is distinctly yellow. The gribenes, or cracklings, are delicious little morsels of crisped skin.

STORING RENDERED FATS

Many people store their rendered fat at room temperature; I do not. I have tasted both lard and duck fat that have gone bad, and it just isn't worth the risk. The folks who store rendered fat at room temperature claim that that is how it was done before refrigeration. But that observation isn't quite true.

In the days before refrigeration, perishable foods were stored in larders and springhouses. Both were designed to keep foods cold. A springhouse literally had a spring running through it, and foods were kept in crocks in the stream.

LONG-TERM STORAGE OF ANIMAL FATS

How long to store animal fats is a remarkably contentious question. There is very little in the way of research, but a lot of opinion.

Rendered fat is a remarkably stable product. Because it is so moisture-free, the normal food spoilage bacteria cannot survive in it. The surfaces that meet the air can become moldy if you use a dirty spoon to scoop it out, but that mold can be scraped off and the rest of the fat used.

All fats will eventually become rancid. The best way to keep your fat from going rancid is to store it in the freezer, where it will keep indefinitely.

A refrigerator will keep an animal fat from going rancid for several months — up to 6 months for lard and tallow, 3 to 4 months for poultry fat. The harder the fat, the longer it should keep. So goose fat will keep longer than duck fat, which will keep longer than chicken fat. Some sources say that tallow will keep indefinitely at room temperature; some claim that lard will also keep indefinitely at room temperature. Since I have tasted lard that has become rancid, I don't trust those claims. If you are storing rendered fat for a coming apocalypse, at least keep it in a dark, cool place, like a cupboard in the cellar.

Fat that is rendered from cooked dishes, such as the fat scraped off a pot of stock or broth, will contain miniscule amounts of the cooked dish, which makes the fat prone to spoilage. Freeze or use such fat within 5 days.

The temperature in a springhouse generally remained in the 50° to 55°F range. The larder was a room where meat was kept under a cap of lard (hence the name) and was expected to remain cold enough for the storage of fresh milk. The larder was usually placed on the north or west side of the house, where it received the least amount of sun. So "room temperature" was and should always be a very cool room temperature.

In urban areas, people didn't stock up with lots of food. They were able to buy small bits of fat as needed. A Chinese friend of mine recalls that his mother used to render a little lard every few days, and his home always had that distinctive odor.

The recommended storage time for all the fats is directly related to the hardness of the fat at room temperature. Tallow keeps longer than leaf lard, which keeps longer than poultry fats and bacon grease.

KOSHER SALT

CURING SALT

CURING FATS

Salting preserves food, fats included. Chances are our ancestors first salted meat — and fats — just to preserve them, then they found that salting transformed the flavor of those foods, improving them. They created the art of charcuterie or *salumi*, along with hundreds of recipes and techniques. Several fatty cuts of pork, including the belly or side meat, fatback, and jowls, can be cured.

The salt used to cure fats (and meats) can be regular sodium chloride in the form of kosher salt or sea salt, or it can be a curing salt that has nitrites mixed in. Kosher salt is preferred because the shape of its flakes are flat, allowing them to cling easily to the meat or fat. Table salt should not used because it contains additives — and people can taste the iodine that is often added to table salt.

The nitrites in curing salts add flavor, preserve the meat's rosy color, prevent the fat from developing rancidity, and inhibit undesirable bacteria from growing. Since nitrites are toxic in large quantities, salts with added nitrites are always dyed pink. Curing salts are found under the general names or brand names of pink curing salt, Prague Powder #1, InstaCure #1, sel rose, Quick Cure, tinted curing mixture (TCM), Modern Cure, DC Cure, and DQ Cure and are readily available online. Pink Himalayan salt is not the same thing. For the recipes here, you can use kosher salt if you don't want to use curing salts.

During curing, salt in a brine or dry cure draws moisture from the fat and creates an environment that is hostile to spoilage bacteria. Herbs and spices might add flavor; sugar may be added to balance out the salt. A cured piece of meat or fat will be firm enough to slice thinly and neatly.

After curing, air-drying or smoking, both of which are beyond the scope of this book, may follow. Air-drying requires a cool, dark, humid space where the temperature is about 60°F and the humidity is about 70 percent. If you are interested in air-drying to make lardo, pancetta, prosciutto, or guanciale, please consult a reputable source, such as Michael Ruhlman and Brian Polcyn's excellent book *Charcuterie*.

CURING SALTS AND CANCER RISK

When you start adding nitrites to meats or fats, people get uneasy. In the 1970s, there was a big question about whether cancer rates could be linked to nitrites added to food. Several studies effectively showed there was no link, but people still stay away from foods with added nitrites, and food processers have gone along with consumer doubt. Today one can buy supposedly uncured bacon and nitrite- or nitrate-free hot dogs. These products are processed using celery juice or beet juice, substances that "naturally" contain more nitrites than traditionally cured fats or meats.

According to the American Meat Institute, roughly 93 percent of our daily intake of nitrites comes from leafy vegetables and tubers. The maximum amount of nitrites allowed in cured meats and fats by the USDA is 156 parts per million (ppm) and is usually lower than that. By contrast, spinach, lettuce, celery, beets, radishes, and carrots can contain up to 1,900 ppm! So when you buy "nitrite-free" or "uncured" or "natural" bacon, you may be getting a bigger dose of nitrites than you would think.

Types of Cured Fats

SALT PORK. Salt pork made from fatback is pure fat. It can also be made from pork belly (the same cut used for bacon), in which case the salt pork will contain layers of both meat and fat, and may be called side meat. Salt pork is usually crisped in a skillet or saucepan until the fat is rendered out; most recipes call for the salt pork to be diced rather than cut into slices. The crisped little bits that are left after the fat is rendered out may be called scrunchions.

PORK BELLY AND JOWLS. Bacon is made from pork belly. Often the pork belly is squared and the trimmings are made into salt pork. Bacon is always cured and then partially cooked by roasting or smoking. Pancetta is pork belly that is cured like bacon, but then it is rolled into a log and hung to dry for a couple of weeks. Pork jowls are almost as fatty as pork bellies and are traditionally made into guanciale, a cured and air-dried product.

FATBACK OR LARDO. Fatback is sometimes cured with salt and herbs. In Italy it is known as lardo. Similar preparations are found throughout Eastern Europe and are called *salo* or *slanina*. A specialty of the Tuscany region of Italy, *lardo di Colonnata*, perhaps the most prized of all cured fats, is fatback that has been cured in a marble box (excluding both air and light) with a mixture of sea salt, ground black pepper, fresh rosemary, and chopped garlic. Lardo is aged for 6 months to 2 years and improves with age. You can often find it served on a charcuterie plate, where it will be sliced thinly and should melt in your mouth, leaving a delicately sweet, herbal creaminess on your tongue.

Lardo is very expensive; it's sometimes available from specialty food stores and in some upscale Italian restaurants in America, but it's a good idea to make sure you're getting the real thing before you pay for it. Lardo is best when made from pigs that are at least 2 years old and have never been fed corn; young pigs or the wrong diet will not yield proper-tasting lardo. Beyond the 2- to 24-week brining period, the fat should be hung to finish curing for at least 4 weeks (preferably about 2 years) in a dark place that is between 45° and 60°F, with between 65 and 75 percent humidity. Although you can find recipes online for making lardo, most are Americanized and simplified, omitting any information about taking care to source the right fatback and often omitting the step of air-drying.

GUANCIALE

BACON

LARDO

SALT PORK

PANCETTA

THE BEST WAY
TO COOK BACON

If you want to save your bacon grease to cook with, it is important to avoid burning it. The best way to avoid scorching the grease is to bake the bacon. Here are the steps:

Arrange the bacon strips on a sheet pan and place the pan on the center rack of a cold oven. Turn the oven on to 400°F and bake for 20 to 35 minutes, depending on the thickness of the bacon strips and how quickly your oven reaches the target temperature. The bacon is done when it is golden brown and crisp.

Transfer the bacon to a paper towel–lined sheet pan to absorb any excess fat. Pour the hot bacon grease through a strainer lined with a coffee filter into a glass jar. Keep the bacon fat refrigerated.

HOME-CURED BACON

Home-cured bacon is a revelation. Its freshness gives it that extra quality, which you won't find even in artisan bacon you might buy. This recipe, which is adapted from Michael Ruhlman's excellent book *Charcuterie*, requires the use of pink curing salt #1. You can get by with all kosher or sea salt, but the bacon will look like well-cooked pork (grayish), taste more like salty pork than like bacon, and will keep for only a week in the fridge. You can easily double this recipe, and freeze whatever you won't eat within two weeks. But the recipe is so easy, I have no problem working in small batches — and besides, there is so much one can do with a pork belly that it is hard to turn it all into bacon.

Makes about 2¼ pounds

   2 tablespoons kosher salt
   1 teaspoon pink curing salt #1
   2 tablespoons coarsely ground black pepper
   1¼ teaspoons Chinese five-spice powder (or substitute other sweet ground spices, such as cinnamon, coriander, mace, cardamom, allspice)
   2 tablespoons honey, maple syrup, or brown sugar
   3 garlic cloves, minced
   1 (2½-pound) piece pork belly, trimmed to a square or rectangle

1. Mix together the kosher salt, pink curing salt, black pepper, five-spice powder, honey, and garlic in a small bowl. Rub the salt and spice mixture all over the meat.

2. Put the pork belly in a gallon-size resealable plastic bag, on a sheet pan, or in a plastic container. Close the bag or cover the meat with plastic wrap and refrigerate for 7 days, turning it over and rubbing the spices into the meat midway through the week.

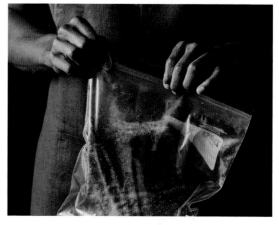

Recipe continues on next page

Home-Cured Bacon, continued

3. Preheat the oven to 200°F or prepare a hot smoker. Remove the bacon from the refrigerator, rinse off all the seasonings under cold water, and pat it dry. Put the meat on a wire rack set on a sheet pan.

4. Roast for 1½ hours or hot-smoke for 4 to 6 hours, until the meat reaches an internal temperature of 150°F. Allow to cool to room temperature.

5. Wrap the bacon tightly in plastic wrap and refrigerate overnight. Slice, then cook in a pan or oven like any other bacon. If the bacon is difficult to slice, place it in the freezer for 1 or 2 hours, until the meat is firm but not frozen hard.

BARBARA PLEASANT'S TALKING BACON

Barbara Pleasant, author of many books, including most recently *Homegrown Pantry*, says she titled this "talking bacon" because every time she walks by the refrigerator, the bacon calls out to her — it is that good. Barbara uses plenty of herbs — fennel, garlic, thyme, black pepper — which are natural preservatives. She speculates that the acids in the coffee must do something, too. She tried switching the syrup to honey, but the molasses makes a better bacon. Orange peel pushed it up another notch.

Makes 1¾–2¾ pounds

- 2 tablespoons fine sea salt
- 2 tablespoons light or dark (not blackstrap) molasses
- 2 tablespoons brewed coffee
- 2 garlic cloves, thinly sliced
- 1 tablespoon coarsely ground black pepper
- 1 teaspoon dried thyme
- 1 teaspoon fennel seeds, crushed
- 1 teaspoon coriander seeds, crushed
- 2 teaspoons finely grated orange zest
- 1 (2- to 3-pound) slab pork belly

1. Mix together the salt, molasses, coffee, garlic, black pepper, thyme, fennel seeds, coriander seeds, and orange zest in a small bowl. Rub the salt and spice mixture all over the belly.

2. Put the pork belly in a gallon-size resealable plastic bag, on a sheet pan, or in a plastic container. Close the bag or cover the meat with plastic wrap and refrigerate for 6 days, turning it over and rubbing the spices into the meat midway through the week.

3. Remove the bacon from the refrigerator, rinse off all the seasonings under cold water, and pat it dry. Put the meat on a wire rack set on a sheet pan. Cover the meat loosely with a dry paper towel. Dry the meat in the refrigerator for 2 days, turning every 12 hours or so.

4. Preheat the oven to 225°F or prepare a hot smoker. Roast for about 1½ hours or hot-smoke for 4 to 6 hours, until the meat reaches an internal temperature of 150°F. Allow to cool to room temperature.

5. Wrap the bacon tightly in plastic wrap and refrigerate overnight. Slice, then cook in a pan or oven like any other bacon. If the bacon is difficult to slice, place it in the freezer for 1 or 2 hours, until the meat is firm but not frozen hard.

SALT PORK

Salt pork is salted and cured pork fat. Traditionally, it was made from the trimmings after the belly was squared for making bacon and from fatback. Though you often see salt pork recipes today calling for a slab of pork belly, I think that cut is too precious to waste on salt pork.

Makes about 2¼ pounds

> 2 tablespoons kosher salt
> 1 teaspoon pink curing salt #1
> 2½ pounds fatback or pork belly trimmings

1. Mix together the kosher salt and pink curing salt in a small bowl.

2. Put the pork in a large gallon-size resealable plastic bag or in a plastic tub with a lid. Rub the salt mixture all over the pork. Close the bag or cover the tub with the lid and refrigerate for 7 days, turning the fat over and rubbing the salt into it midway through the week.

3. Use within a month; or cut into pieces that will be used up in a month, wrap well, and freeze for long-term storage. Salt pork picks up freezer odors easily, so make sure each piece is well wrapped.

Cure salt pork with or without the skin. This fatback has the skin still attached.

SALT PORK

In days gone by, farmers and homesteaders made salt pork with every piece of meat that wasn't eaten immediately, smoked, or turned into sausage. They would put whole sides of pork, or just parts of the pork carcass, into a barrel and cover it with salty brine. Such barrels of pork are the origins of the phrase "scraping the bottom of a barrel" (when most of the pork had been consumed) and also "pork barrel legislation," which gives funding for special projects based on political favors rather than on need. The pork barrel was ubiquitous in the United States from the Revolutionary War to the Civil War.

Salt pork is the signature fat in New England–style chowders. Old-timers may remember suppers of salt pork and milk gravy, served over biscuits. Ma made it in *Little House on the Prairie*; a single restaurant — the Wayside Restaurant and Bakery in Montpelier — keeps the tradition alive in Vermont. The dish is the Northern version of the famous Southern breakfast of sausage gravy and biscuits (see page 166).

BETTER BUTTER

Butter is most everyone's favorite fat. We grew up eating it, we are familiar with baking and cooking with it. So I haven't devoted much space to it in this book — you can find plenty of recipes and information on butter elsewhere.

Compared to the animal fats covered in this book, butter is higher in saturated fat (50 percent compared to 39 percent of lard) and lower in monounsaturated fat (30 percent compared to 45 percent of lard). Unlike other animal fats, which are rendered to remove everything but the fat, butter is made up of about 80 percent fat and 20 percent water and milk solids. It has a relatively low smoke point (350°F), making it unsuited for frying, and it has a melting temperature of 90° to 98.6°F. But it does taste good and makes lovely baked goods. Feel free to substitute half butter for any baking recipe that calls for lard and you should have fine results.

Butter at its best and freshest reflects the cows' diet — and butter from pasture-raised cows is high in conjugated linoleic acid (CLA), which acts like an omega-3 fatty acid and supposedly protects against heart disease and inflammation.

Homemade butter often tastes better than the standard American store-bought butter. Better butter starts with cultured cream. Not surprisingly, this was the American standard not so long ago. Typically, the cream was left out all night to clabber. Clabbered milk is milk that has thickened slightly and ripened in flavor. The flavor is pleasantly rich, nutty, slightly tangy but not at all sour.

Once agriculture was industrialized and refrigeration was introduced, butter making moved off the farm and into creameries. These creameries made "sweet butter" from fresh cream, in part because it speeded up the process. Consumers grew used to its mild, unctuous flavor.

Sweet cream butter became the standard American butter, the butter most of us were raised on. Although salt was once added as a preservative, refrigeration made that unnecessary. Salted and unsalted butter, now sometimes called sweet butter, became the standards. But it is all sweet (in the original meaning of not cultured) butter.

What was lost in the switch from cultured to sweet butter? Only the flavor. Naturally occurring lactobacilli ripen fresh cream into cultured cream. When the thickened cream is churned into butter, it is rich and flavorful, as opposed to unctuous, smooth, and neutral tasting.

You can make cultured butter from store-bought pasteurized cream, but not from ultra-pasteurized cream. Read the labels carefully. My only source for pasteurized, not ultra-pasteurized, cream is from a local dairy. All the national brands are ultra-pasteurized. If you are using pasteurized cream, add cultured buttermilk (use 2 tablespoons buttermilk to 2 cups cream). Bring the cream to 68° to 70°F (do not let it fall below 68°F or rise above 78°F), add the culture, and keep covered and warm for the next 6 to 12 hours, depending on how strongly cultured you want your butter. Alternatively, you can add a culture purchased from a cheese-making supply store. Choose a buttermilk culture.

After this ripening, the cream should be noticeably thicker and have a well-developed aroma. It should taste delicious, slightly sour, and have no after-taste. If the cream is bubbly, or smells yeasty, gassy, or off in any way, you have a contamination problem: Throw the cream away and start over.

Wrap butter in parchment paper or waxed paper and roll on the counter to form an evenly shaped roll.

MAKING CULTURED BUTTER

You can make butter in a stand mixer, a food processor, a blender, or an immersion blender in a mason jar.

Start with raw milk and let it stand overnight in the refrigerator until the cream rises to the top. Skim off the cream. Return the milk to the refrigerator but let just the cream stand at room temperature for 12 to 24 hours. The flavor will ripen and the cream will thicken slightly. If you are using pasteurized cream, add 2 tablespoons buttermilk to each 2 cups heavy cream (or use a buttermilk culture) and let stand for 12 hours at room temperature.

Pour the cream into the bowl of a stand mixer fitted with the whip attachment. Beat on medium-low speed until you have whipped cream. Continue beating until the cream separates into globules of fat and a milky liquid. Stop beating immediately (don't overbeat).

Strain out the butterfat (the milky white solids) over a bowl. You'll want to keep the buttermilk (what drains out) for drinking, making a creamy salad dressing, or using in pancakes or biscuits. The flavor of fresh buttermilk is delicious, which shouldn't come as a surprise, but always does. Hold the strainer with the butter under running water until the rinse water runs clear.

Briefly knead the butter with two wooden spoons to squeeze out any remaining moisture. This step is really important for the flavor of the butter. Any remaining buttermilk will sour and give the butter an off flavor. Fold in salt, if you like. I prefer to add a flake salt, like Maldon sea salt. Use about ½ teaspoon salt for 2 cups cream.

The yield from 2 cups cream is about 1 cup of butter. Butter should be kept in a covered dish and will keep in the refrigerator for several weeks. Excess butter can be stored in the freezer for up to 9 months; double-wrap in plastic wrap before bagging or wrapping in freezer paper, or use a vacuum sealer; butter picks up off odors in the freezer.

Knead the butter
to press out
any remaining
buttermilk.

CHAPTER 3

Cooking and Baking with Animal Fats

*C*ooks have been turning out delicious food based on animal fats far longer than they have been cooking with any vegetable seed oil, and also far longer than they have been cooking with olive oil outside of olive-growing regions. So if you browse the Internet for traditional recipes, you'll find many of them were originally made with animal fats, particularly lard. Why lard? Because it was abundant and cheap, and produced great results. While pigs can be raised just about anywhere people live, olives for olive oil are found in a relatively small geographic area. Even in olive-growing regions, where pork wasn't prohibited by religion, cooks were just as likely to use lard for cooking and baking as they were olive oil. Butter, even in northern dairy regions, was always more limited in supply than lard.

COOKING WITH ANIMAL FATS

Birds produce less fat than pigs or cows do, but poultry fats are equally wonderful to cook with. Traditional recipes using poultry fats are limited mainly to Jewish, German, and French cuisine. Jews, of course, were prohibited from cooking with pork, but many of their dishes in Eastern Europe were quite similar to dishes non-Jewish Eastern Europeans prepared, just made with goose fat instead of lard. (The chicken fat came later, when Jews immigrated to the United States, settled in urban areas, and found goose fat in short supply.) The French raise a lot of ducks and geese, and they make abundant use of these fats, particularly in confit dishes (where poultry is immersed in fat and slowly simmered in the oven). In German cuisine, schmaltz is both rendered goose fat and rendered pork fat; ducks are less common than geese.

The biggest surprise that comes with cooking with animal fats is how they enhance flavor without adding meaty notes. This may sound like a contradiction, but it is true. As proof of concept, please cook your morning eggs or your evening popcorn in rendered duck fat and see how the flavor is heightened. The eggs will taste richer, but not ducky. The popcorn will taste like popcorn should and won't need any butter. Fat carries flavor, and animal fats do a very good job of it.

Baked-on vegetable seed oils cause a gummy buildup on ovenware and cast iron. Animal fats, on the other hand, keep cast iron well seasoned and don't cause a sticky residue on any type of cookware. When fast-food joints switched from tallow to vegetable oils to make French fries, they were forced to deal with the gummy, sticky, flammable residue that built up in their deep fryers. Flammable residue in the clothing of the workers was and is another problem, as are carcinogenic fumes inhaled by the workers. None of this occurs with animal fats.

When it comes to cooking with animal fats on top of the stove, you should find it requires little in the way of special knowledge compared to cooking with vegetable seed oils or butter. Baking with animal fats is different, but only somewhat.

THE JOY OF FATS

One of the reasons why fat is so important in cooking is that fats conduct heat without boiling away like water will. This allows meat to take on a crusty sear — the Maillard reaction — at temperatures above 300°F. Fats also lubricate food and prevent them from sticking to a hot pan. Fats both provide flavor and carry the flavor of fat-soluble volatile flavoring compounds, like the volatile oils that make herbs so potent.

Animal fats allow foods to taste like themselves, only better. There's a reason so many restaurants are using duck fat to make French fries, or that McDonald's originally used tallow in their deep-fat fryers. Bring the tallow back! It is a great frying medium.

Animal fats are simply superior when it comes to browning foods on top of the stove, especially compared to cooking with butter, which is an emulsion of fat, water, and milk solids. When butter is heated, it breaks down and the water is released, creating steam. The effect is subtle, but it does slow the browning action. Also, butter has a comparatively low smoke point, so when sautéing, most chefs prefer to use half butter for flavor and half seed oil to raise the smoke point; half butter and half other animal fats will provide the same effect: buttery flavor, high smoke point.

MEASURING ANIMAL FATS

I have adopted the practice of weighing fats when I cook, and I use gram measures. Therefore, the gram weights are the most accurate measures you can use in this book. In America, most of us have been taught to cook using volume measures. These volume measures (spoons and cups) were standardized by Fannie Farmer, who made it her mission to standardize recipes that called for a handful of this and a teacup of that. So now, an American cook might make a pound cake based on 3½ cups of flour and 1 cup of butter.

Then the rest of the world adopted the metric system and recipes in the rest of the world came to be written by weighing out the solids in grams and using milliliters for liquids. So for that same pound cake recipe, an English cook would reach for a scale and measure out 440 grams of flour and 225 grams of butter — even though 3½ cups of flour weighs 437.5 grams and 1 cup of butter actually weighs 227 grams. In other words, the gram-to-cup equivalents are rarely exact; they are usually rounded up or down, because people don't like fussing when measuring.

I measure animal fats by weight on a digital scale. Gram weights are the most accurate.

Although I have double-checked my lard measures frequently (1 packed cup equals 200 grams), there is still a problem with ounce weights: 200 grams equals 7.94 ounces, which we round up to 8 ounces. Also, it takes a fair amount of mashing and pressing to be sure that 1 cup of measured lard doesn't contain any air holes. All that handling softens the lard, which is less than ideal when making pie pastry. Weighing it out is just easier.

Tallow is tricky to measure because it is hard and brittle. In order to measure it by volume, you have to grate it. Then you have to really, really mash it into a cup to get rid of any air that might exist among the shreds of grated fat. Your measure of tallow may not be accurate by volume; it is far easier to just weigh it, breaking off pieces to get to the right weight.

When measuring out poultry fats, there is no problem measuring with spoons or cups when the fat is at room temperature or when chilled because it is liquid at room temperature and still very, very soft straight out of the fridge. Still, I find it easier to weigh it out rather than fuss with leveling the spoons or cups.

STOVETOP COOKING

On top of the stove, you generally sauté, stir-fry, panfry, shallow-fry, and deep-fry — and make popcorn. For sautéing and stir-frying, you generally want to coat the bottom of your pan or wok with fat. With both stir-frying and sautéing, you want to keep the foods moving and the heat high. Because animal fats are not absorbed as readily as vegetable seed oils and because they do a better job at creating a nonstick surface, you may find you need less fat than you are used to when sautéing or stir-frying. Other than that, you won't find much difference from sautéing or stir-frying with a vegetable seed oil. In general, it doesn't matter which animal fat you choose to work with here, though lard and poultry fats tend to enhance flavor more than tallow, and bacon fat and salt pork add flavor. Bacon sometimes adds too much flavor, so use it sparingly or in combination with another fat.

HIGH–TEMPERATURE COOKING AND FAT SMOKE POINTS

It is important to know the smoke point of a fat and to keep the temperature of the fat below that point. Once a fat starts to smoke, it usually will emit a harsh smell and fill the air with smoke. Fats that have gone past their smoke points contain a large quantity of free radicals, which contribute to the risk of cancer. High-temperature cooking includes pan-searing, stir-frying, sautéing, shallow-frying, and deep-frying. Fortunately, all the animal fats are up to the job.

FAT	MELTING POINT	SMOKE POINT	CHARACTERISTICS
Beef tallow	115° to 120°F	375° to 400°F	White, brittle
Butter	90° to 98.6°F	350°F	Yellow, soft at room temperature
Chicken fat	73° to 104°F	375°F	Yellow, semisolid
Duck fat	78° to 110°F	375°F	Yellow, semisolid
Goose fat	80° to 110°F	375°F	White, semisolid
Lamb, goat tallow	122°F	370°F	White, hard
Lard	110° to 112°F	370°F	White, hard

Shallow-Frying or Panfrying

Here's where animal fats begin to show how much they improve your cooking. Food fried in animal fats have a far less oily-tasting surface than food fried in vegetable seed oils, because the saturated fat solidifies as it cools and cannot be absorbed into the food. Plus, some fats, like all poultry fats, are more flavorful than oils. So from a flavor and texture standpoint, animal fats are best. As with sautéing and stir-frying, it doesn't matter which animal fat you choose to work with, though, again, poultry fats tend to enhance flavor more than lard or tallow,

RIGHT: Home Fries (page 204)

and bacon fat and salt pork add flavor. Again, bacon can add too much flavor, so use it sparingly or in combination with another fat.

Animal fats are also superior when it comes to even browning and preventing the food from sticking to the pan. Just try home fries in lard (page 204) and marvel at how evenly the potatoes brown, without sticking. Or fry tofu in any animal fat and note how evenly the cubes brown, how none sticks to the pan.

When shallow-frying, how much fat to use depends on the recipe in question and what you are frying. You can use as little as ¼ inch or as much as 1 inch. You want the fat hot, but not smoking. It can be difficult to gauge the temperature of the fat when shallow-frying because it is hard to keep the tip of the thermometer in the fat and off the bottom of the pan. Here, the bread cube test is handy: A bread cube will brown in 60 seconds when the fat is at 350°F, a good frying temperature. Food releases more easily from the pan and browns more readily when frying in animal fat than in a vegetable seed oil. Usually, you will want to flip the food only once.

A cast-iron skillet is perfect for shallow-frying or panfrying because animal fats are perfect for seasoning them (see page 127). After each use, just wipe out the skillet and give it a quick rinse in hot soapy water. There should be no sticking with a well-seasoned cast-iron skillet.

If you need to hold the foods after panfrying or shallow-frying, hold them in a 200°F oven on wire racks set over sheet pans to retain the crisp texture. (I prefer wire racks over paper towels because they keep the cooked food crispier.)

Deep-Frying

Deep-frying evaporates the moisture from your food and causes it to release steam (that's the bubbling you see around each piece of food in the hot fat). The outflow of that steam prevents the fat from seeping in. As long as you do a good job of keeping your fat hot enough, your foods will emerge from the hot fat crisp and not greasy. They will hold up for enjoying as leftovers, remaining relatively crisp and completely not greasy. Despite their unhealthy reputation,

RIGHT: Korean Fried Chicken Wings (page 122)

TIPS FOR DEEP FRYING

Here are some tips to guarantee success with deep-frying:

- Adding ½ teaspoon of baking powder per ½ cup flour will make batter coatings lighter.

- After breading foods that are to be fried, let them air-dry at room temperature for about 30 minutes to reduce spatters.

- When deep-frying in a wok with a rounded bottom, be sure the wok is set on the wok ring to remain stable.

- Keep an eye on the fat as it heats. You don't want it to start smoking, which means that it is starting to break down.

- Don't add too many items at a time or the temperature of the fat will drop too much.

- Between batches of food, skim out stray bits of breading or batter and discard. If you leave them in, they will burn and contribute a burnt flavor to the fat.

- Remove the cooked food from the hot fat and drain on wire racks set over a sheet pan to catch drips. You can augment the wire racks by gently patting the surface of the fried foods with paper towels.

- The temperature of the fat will drop as soon as the food to be fried is added. If you want to fry at 350°F, you will need to heat the fat to 375°F. Once the food has been added, try to keep the temperature as close to 350°F as you can by adjusting the burner or the thermostat on the fryer.

deep-fried foods usually contain no more fat than stir-fried or sautéed dishes as long as they were cooked in fat that was hot enough.

Deep-frying at home has its challenges. It requires at least 4 cups (3 pounds/800g) of rendered fat, and cleanup is time-consuming. But — and here's the important *but* — if you are a fan of doughnuts, of fried chicken, of fried *anything* — it's better to fry in lard or tallow than in vegetable oil for really good results.

ELECTRIC FRYER OR SKILLET?

A lot of people swear by their electric deep fryers — but there's a catch with the newer models; most have a filtering mechanism that may not work with solid fats. So it is necessary to make sure it can handle a solid fat before buying this appliance. Most are meant to be cleaned by first draining out the cooled liquid oil; animal fats will not drain unless they are warm enough to be melted (110° to 112°F for lard; 115° to 120°F for tallow). If you fry something for dinner, the fat will be solid by the time you are ready to clean up. So you must reheat the fat and catch it at the very right temperature to transfer it into a container that won't melt or break. And, of course, every time you heat a fat, you lower the smoke point and bring it closer to rancidity.

Another catch with electric deep fryers is that you may not be able to control the temperature of the fat very well because some have no temperature controls, but rather just on-and-off switches. Price matters here. Read online reviews before deciding on a model to avoid disappointment.

When deep-frying in a saucepan, deep skillet, or wok, you can use less fat and turn your foods over rather than completely submerge them. With a deep fryer, you need to add as much fat as your appliance requires — and it is often more than what is required in a small saucepan. This is great for large batches but less great for conserving your fat.

STORING AND REUSING FAT

Storage of the used fat is another issue. Provided that fat has not reached its smoking temperature (see table, page 76), fats used for deep-frying can be reused after straining. If you use your fryer on a daily basis, you can probably keep the fat in the fryer and reuse it until it begins to darken or smell off. Otherwise, after each use, drain it from the deep fryer, strain it, and store the fat in a cold, dark place (the refrigerator or freezer is best). How many times you can reuse the fat will vary, but use the color as a guide; it will go from white to beige to brown. Discard it when it is brown or smells off. And definitely don't expect the fat in which you fried fish to taste fine with other foods.

Many advise reusing fat only three times. This is because the smoke point of the fat lowers each time it is heated, and as the fat begins to oxidize, it begins to be rancid. It also becomes flammable at a lower temperature. Unless you are frying a naked food — like a French fry or veggie chip — bits of breading and batter will need to be strained from the fat before it can be reused. Otherwise the debris will scorch and flavor the fat.

Strain fat used for frying. If it's nearly white when hardened, and odor free, it can be used again.

WHICH FAT TO USE

There are absolutely no rules about which fat to use when deep-frying; they all do a fine job, but lard and tallow are easier and less expensive to acquire in quantities needed for frying than poultry fats. They are also more neutral in flavor. Tallow has the highest smoke point of 400°F, which is the same as canola oil. Poultry fats have a smoke point of 375°F, while lard comes in at 370°F. Remember that fats start to break down and become rancid more quickly once they have reached their smoke point. Bacon grease is inappropriate for deep-frying because its smoke point has already been lowered to 325°F from 370°F (lard's smoke point).

TEMPERATURE MATTERS!

Your success with deep-frying greatly depends on the temperature of the oil, and for that you will need a candy or deep-fry thermometer, if you are doing your deep-frying in a saucepan or skillet. Yes, you can judge that a bread cube will brown in 60 seconds when the fat is at 350°F, but what is going to tell you how much the temperature has dropped once the food has been added? And how do you measure a temperature of 320°F, which is needed for the first, blanching step when making French fries? A thermometer is necessary to regulate the temperature throughout the process.

There is plenty of chatter on the Internet about the accuracy of the thermostats on various brands of electric deep fryers. So even if you use one of these deep fryers, it is still a good idea to confirm the temperature with a thermometer.

Another very useful tool is the spider — a long-handled tool with a wide, flat wire mesh basket. It does an excellent job of lifting up fried foods while straining out the hot fat; it has more straining ability than either a slotted spoon or slotted spatula. And it comes in handy for removing small bits of debris, even if you use a fry basket. The least expensive ones have bamboo handles, which are better than plastic handles, which can melt if left in the pot.

For the best results, monitor the temperature of your fats when frying.

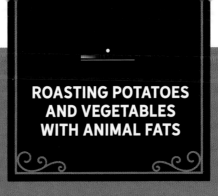

ROASTING POTATOES AND VEGETABLES WITH ANIMAL FATS

The trick here is to preheat the half sheet pan or shallow roasting pan with the oven, then add three-quarters of the fat you usually use (for example, 3 tablespoons instead of 4 tablespoons) and the fat will instantly melt. Turn and tip the pan so the bottom is evenly covered with the fat, then use two silicone spatulas to toss the veggies in the pan until they are evenly coated with the fat. Then roast. I don't recommend using bacon grease here because of its relatively low smoke point and strong flavor that can mask the delicate flavor of vegetables.

Any recipe or technique that requires oil can be adapted to use animal fats as long as the dish is baked or served hot. Here are a few ideas.

ROASTING VEGETABLES: Place a sheet pan in the oven and preheat to 425°F. When the pan is heated, add the fat, using three-quarters of the amount you usually use. Turn and tilt the pan until the fat is melted. Add the veggies and toss with two silicone spatulas to get the veggies evenly coated. Roast, turning or shaking the pan as the veggies roast. Poultry fats add extra flavor, but any animal fat can be used. Roasting root vegetables and potatoes in any poultry fat is particularly scrumptious.

FRIED VEGETABLE CHIPS: Slice any root vegetable on a mandoline or shred with a spiralizer, deep-fry at 350°F, drain, then sprinkle with salt.

ANY POTATO RECIPES: Poultry fat can replace butter in mashed potatoes; it is also wonderful with roasted potatoes. Panfry for nongreasy pancakes, hash browns, or home fries. Tallow is excellent for deep-frying.

SAUTÉING VEGETABLES: Use about three-quarters of the amount of fat you usually use. Sauté in poultry fat over high heat for extra flavor.

EGGS ANY WHICH WAY: Use poultry fats instead of butter for the richest flavor.

RICE PILAFS AND RISOTTOS: Substitute any animal fat. Poultry fats add a depth of flavor and complement any recipe that uses chicken broth as the cooking liquid. Mexican (or Spanish) rice should be made with lard.

STIR-FRIES: Lard and tallow withstand the high heat of wok cooking and make dishes that are not greasy. Before the 1980s, lard was the most common cooking fat for stir-fries in China.

PAN-SEARED MEATS AND FISH: Match the meats with the animal fats. Use poultry fats or bacon grease with fish.

PIE PASTRY: Short-crust pastry can be made with any animal fat. Leaf lard pastries handle like a dream. Tallow requires grating or chopping with some flour in a food processor to get the texture fine enough to mix with the flour. It makes pastries as flaky as those made with lard. Pastries made with poultry fats can be a little harder to handle than those made with lard or tallow, and the pastry dough will be softer and less rigid. Chill pastry made with poultry fats until firm before using it.

QUICK BREADS: Substitute melted fat for the oil the recipe calls for.

BAKING WITH ANIMAL FATS

The misconception that people have about baking with animal fats begins and ends with the idea that the animal fats impart a meaty flavor. Properly rendered lard and tallow simply do not have a meaty flavor. Poultry fats do have a detectable flavor and are softer than tallow and lard, and that does limit their use somewhat. That said, I've made piecrust for sweet pies with chicken fat and no one noticed any odd flavors — and it was marvelously crisp.

Animal fats work much the same way butter (also an animal fat) does in baking. The fat interferes with the development of gluten by "shortening" the protein strands that are created when flour is mixed with water; this tenderizes a dough. Fat conducts heat through a batter to aid in browning. When beaten, the fat holds air to lighten a batter or frosting, though not as fully as butter does. Finally, fat provides the illusion of moisture or wetness on the palate.

The fat crystals in animal fats are also larger than those in butter, which means there is more empty space left behind when the fat melts — more empty space also means more layers and flakes.

Animal fats are 100 percent fat, just like vegetable shortening, whereas butter is about 80 percent fat, 15 percent water, and 5 percent milk solids. You can substitute any animal fat in any recipe that calls for shortening without making any changes to the recipe. If you are substituting an animal fat for butter, you will sometimes have to compensate for the absence of either water or milk sugars or both.

When it comes to greasing the bakeware, lard is most easily applied with a paper towel in much the same way butter is used. It will offer superior nonstick qualities and will not cause a buildup of a gummy finish. Poultry fat and bacon grease can also be used, though bacon grease will definitely contribute flavor. This may be exactly what you want in a corn bread, but not in a pound cake.

Baked goods made with bacon grease tend to have a greasy mouthfeel, and the bacon flavor is too pronounced (in my opinion). If you want to experiment with bacon grease, try using just a tablespoon or two and using lard for the remaining fat needed. Consider also that bacon grease will add salt and sugar, further affecting the flavor of whatever it is you are baking.

LARD
100% fat

BUTTER
80% fat,
15% water,
5% milk solids

Pastry and Pie Dough

In a pastry or pie dough, the fat distributed in the flour physically prevents the proteins in the flour from bonding with each other and with water to form gluten, which toughens a crust. During baking, the water turns to steam and the fat melts, leaving empty pockets between layers of pastry to give a crust its flaky quality. Many chefs prefer a mixture of lard, for the texture, and butter, for the flavor. This suggests that butter is necessary for the flavor. In fact, an all-lard piecrust tastes just fine. Poultry fats, leaf lard, and tallow all work well here; bacon grease does not.

In most pastry and pie doughs, the absence of water is an advantage. This is the reason pie doughs shrink less when made with an animal fat other than butter. But the absence of milk sugars delays the browning, and overbaking can result in an overly crisp pastry. Also, while butter imparts a buttery, nutty flavor, lard and tallow should be completely neutral and contribute nothing in terms of flavor. Poultry fats add an elusive richness that is almost like butter and does work in a sweet pie.

To adapt recipes for pastry and pie doughs:

- Increase the salt by ¼ teaspoon per cup flour if you usually use unsalted butter or by ½ teaspoon per cup flour if you usually use salted butter.

- Increase or add sugar by 1 tablespoon per cup flour for savory recipes or by 2 to 3 tablespoons per cup flour for sweet ones.

- Avoid overbaking by checking the feel of the pastry as well as the color. An overbaked pastry will shatter when cut.

- Poultry fats are quite soft compared to butter and other animal fats, with chicken fat the most soft, duck fat a little more solid, and goose fat even more solid, at least when very cold or frozen. When making pastries with poultry fats, work with frozen or very, very cold fats for the best results, and chill the pastry before baking.

- When making pastries, use about three-quarters of the fat the butter recipe calls for to avoid a greasy texture.

- Leaf lard is better for texture than lard made from other pork fats.

LEFT: Blueberry Galette (page 241)

Cookies

Many traditional cookie recipes were made with leaf lard in the first place. Cookies made with lard or tallow instead of butter will spread less and will have a crunchier texture. The melting temperature of butter is about 98.6°F, while the melting point of lard is 110° to 112°F and the melting point of tallow is 115° to 120°F. As they bake, butter-based cookies start to spread before their structure is set because the butter melts so quickly. Cookies made with lard or tallow begin to set before the fats' melting points are reached. So cookies made with lard or tallow tend to be taller and chewier whereas butter-based cookies are wider, thinner, and crispier.

An old recipe from your grandmother's recipe box that calls for Crisco will work just fine in a cookie recipe, without changes. But if you want to substitute lard or tallow for butter in a cookie recipe, adapt the recipe with the following suggestions:

- If you are working with tallow, make the dough in a food processor fitted with the steel blade. Grind up the tallow with ½ cup of the measured amount of flour until the mixture is evenly blended and looks like tiny, tiny pebbles.

- Increase the salt by ¼ teaspoon per cup flour if you usually use unsalted butter or by ½ teaspoon per cup flour if you usually use salted butter.

- Increase the sugar by 2 to 3 tablespoons per cup flour for better browning.

- Avoid overbaking by checking the feel of the cookie rather than using color as your sole guide, especially if you haven't increased the sugar. The top of the cookie should be set and not wet.

- For the best results, regardless of the fat used, chill the formed cookies before baking. This may require a few minutes extra of baking time.

- If the cookies seem greasy, use three-quarters of the amount of fat next time (remember that butter is only 80 percent fat).

RIGHT: Crinkle-Top Molasses Cookies (page 228)

Quick Breads and Muffins

These recipes are easily adapted to animal fats because they rely on eggs and baking powder for leavening rather than on creaming the butter and sugar. Look for recipes that are made with melted butter, oil, or vegetable shortening rather than butter; these don't require adapting, other than melting your animal fat.

Cakes

Animal fats do not hold air as well as butter does when creaming with the sugar, so adapting a butter-based recipe presents challenges. Any recipe already adapted to oil should work, as is the case with a lot of carrot cake recipes. Look for recipes that use more than two eggs for leavening. If using tallow, grind it with some of the measured flour to allow it to become fully incorporated into the batter.

You also have Crisco to thank for a lot of recipes adapted to use vegetable shortening. In fact, Procter & Gamble were quite aggressive about marketing their product by giving away free cookbook collections using vegetable shortening in classic cakes. Work with lard or tallow at room temperature for easier blending into the batter. Your cake made with lard or tallow may not rise quite as high as ones made with vegetable shortening or butter.

To compensate for the absence of milk sugars from butter, be sure to use high-quality extracts and other flavoring ingredients like cocoa powder.

Yeast Breads

Yeast breads are basically divided into two types: lean and enriched. The lean versions — think baguettes and many Italian breads — are virtually fat-free. Enriched breads are made with fat, sugar, and eggs, and often the fat was originally lard, though modern recipes usually specify butter. You can replace oil with an equal amount of melted animal fat. Replace butter with three-quarters of the amount of fat and add a little extra salt and sugar.

LEFT: Carrot Cake Cupcakes (page 246)

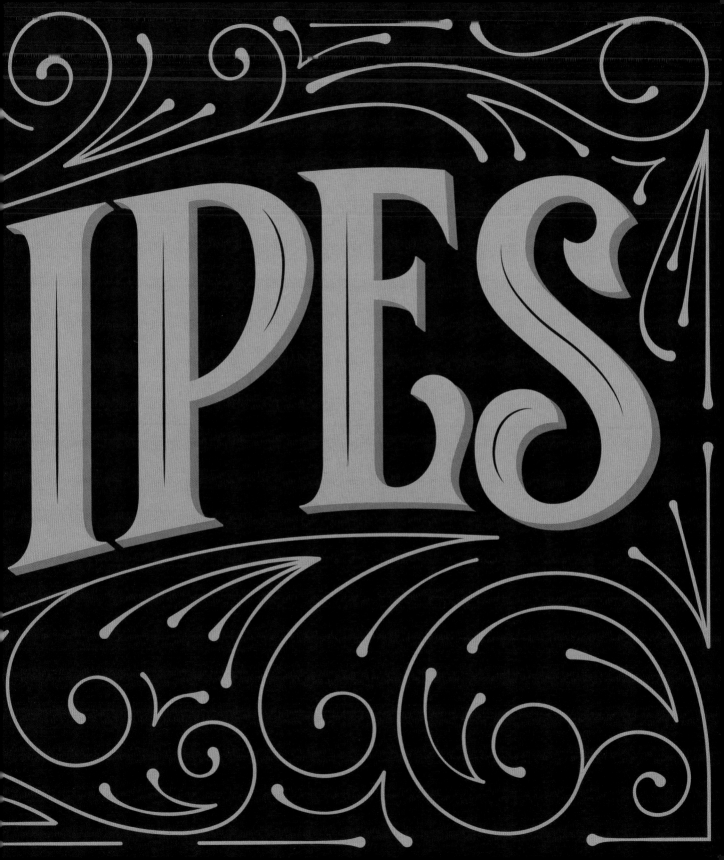

Korean Fried
Chicken Wings,
page 122

CHAPTER 4

Snacks, Street Food, and Starters

CRISPIER KALE CHIPS

Perhaps you already make kale chips. Perhaps you think the recipe can't be improved. Well, it can, with any poultry fat. Kale chips made with oil usually lose their crisp texture after a few hours. Not so with poultry fat. Duck and goose fat rule here, but chicken fat is just fine also. Crispy kale chips are the perfect salty snack when you aren't in the mood for popcorn.

Makes about 8 cups

 1 bunch curly kale, tough stems discarded and leaves chopped into 1-inch pieces
 2 tablespoons (0.8 ounce/25g) any poultry fat, melted
   Coarse or fine sea salt

1. Preheat the oven to 425°F.

2. Put the kale in a large bowl. Pour in the melted fat and mix well with your hands to make sure the leaves are evenly coated. Spread out on a large sheet pan in a single layer.

3. Roast for 10 to 15 minutes, until the curly tips of the leaves are darkened and the center of the leaves are a bright green. The leaves should be mostly crunchy, but not blackened.

4. Toss with salt and serve.

DUCK FAT POPCORN

In truth, any animal fat can be used to make popcorn, but it sounds so luxurious when you add "duck fat" to the title — and poultry fat is my favorite popping medium. How does an animal fat change the flavor of popcorn? It doesn't — it makes it taste right, like popcorn should taste. My husband says, "If popcorn were made in heaven, this is how it would taste." Don't bother with adding butter to this perfect treat.

Makes about 10 cups

 3 tablespoons (1.4 ounces/40g) duck fat
 ⅓ cup popcorn kernels
   Coarse or fine sea salt

1. Melt the fat in a medium saucepan over medium heat. When the fat has melted, add the popcorn kernels, and shake the pan to spread the fat evenly and coat all the kernels. Cover the pan and cook, shaking frequently, until the popping begins to slow (about 2 seconds between pops). Remove the pan from the heat and let it stand while the last few kernels pop.

2. When the popping has stopped completely, remove the lid from the pan, dump into a large bowl, and season the popcorn generously with salt. Toss to distribute the salt and serve immediately.

DUCK FAT MAPLE–CARAMEL POPCORN

This is so seriously addictive that you'd be wise to make it only when you have plenty of other hands dipping into the bowl. Once you start, you won't stop until the last kernel is gone.

Makes about 12 cups

- 3 tablespoons (1.4 ounces/40g) plus
  ½ cup (3.5 ounces/100g) duck fat
  (or any poultry fat)
- ½ cup popcorn kernels
- 1 cup lightly packed brown sugar
- 1 teaspoon fine sea salt
- ¼ cup pure maple syrup
- ¼ teaspoon baking soda

1. To make the popcorn, melt the 3 tablespoons fat in a saucepan over medium heat. Add the popcorn kernels and shake the pan to spread the fat evenly and coat all the kernels. Cover the pan and cook, shaking frequently, until the popping begins to slow (about 2 seconds between pops). Remove the pan from the heat and let it stand while the last few kernels pop.

2. When the popping has stopped completely, remove the lid from the pan and dump the popped corn into a very large bowl — large enough so you can toss the popcorn with the caramel and not have it spill all over the counter.

3. Preheat the oven to 275°F. Line a half sheet pan with parchment paper.

4. Combine the remaining ½ cup fat, the brown sugar, salt, and maple syrup in a saucepan over medium-high heat. Stir gently until the mixture comes to a rolling boil and reaches 230°F on a candy thermometer. Remove the sugar mixture from the heat. Add the baking soda (it will foam up), stir, and pour the mixture over the popcorn. Mix gently with two spoons until well coated.

5. Spread the popcorn out on the lined sheet pan and bake for 45 minutes, stirring every 15 minutes. Remove from the oven and give a final mix on the pan to reincorporate the caramel that has melted off. Spread out on the same baking sheet, give the popcorn a final stir, and let cool. Once cool, enjoy or store in an airtight container for 4 to 7 days.

HOMEMADE POTATO CHIPS

These chips are fried at 325°F instead of the usual deep-fry temperature of 365°F. The lower temperature results in a crispier chip. And because they are fried in animal fat, they remain crisp longer (unless you are making this in a muggy climate, in which case, nothing keeps fried foods crisp for long). A mandoline works best for slicing the potatoes. Any thicker than ⅛ inch and the chips will not be crisp, any thinner and the chips will burn quickly.

Serves 4

 1½ pounds russet potatoes, sliced ⅛ inch thick
       (peeling is optional)
 3½ cups (1½ pounds/700g) lard or tallow
       Coarse or fine sea salt

1. As soon as the potatoes are sliced, put them in a bowl of cold water. Swish the potatoes around to wash the starch off the surface of the potatoes. Lift the potatoes out of the water and put in a colander. Rinse out the bowl and fill with fresh water. Return the potatoes to the water and let sit for at least 30 minutes, or up to 8 hours.

2. Arrange a heavy towel on the counter. Lift the potatoes out of the water and spread out in a single layer on the towel. While the potatoes are drying, melt the lard in a large wok and heat to 325°F. Set out a colander over a plate. Line a large bowl with paper towels. Set out a serving bowl. Roll up the towel with the potato slices or pat the potato slices dry.

3. Add about one-quarter of the potatoes to the hot fat and fry, stirring the chips and flipping them constantly with a wire-mesh spider, until the bubbling slows and the chips are pale golden brown, about 5 minutes.

4. Lift the chips out of the fat with the spider, transfer to the colander, and shake to encourage the fat to drain off. Dump the chips into the paper towel–lined bowl, season with salt, and toss. Allow to drain for 30 seconds, then transfer to the serving bowl. Repeat with the remaining potato slices in three more batches, watching the temperature of the lard to maintain a constant temperature.

5. Serve immediately.

RESTAURANT–STYLE TORTILLA CHIPS

When you go to most Mexican restaurants in the United States, a basket of chips and a small bowl of salsa is brought to your table almost instantly. Those chips are usually made in the restaurant from stale corn tortillas and taste like the ones you can produce with this recipe. If you want to make this in a deep fryer or saucepan, you may need more fat.

Serves 4–6

3½ cups (1½ pounds/700g) lard or tallow
16 fresh or stale corn tortillas, cut into
      6 wedges each
      Coarse or fine sea salt

1. Melt the fat in a large wok and heat to 350°F. Set a colander on a plate to catch drips. Line a large bowl with paper towels. Set out a serving bowl.

2. Add about one-quarter of the tortilla pieces to the hot fat and fry, stirring the chips and flipping them constantly with a wire-mesh spider, until the bubbling slows and the chips are pale golden brown, about 2 minutes.

3. Lift the chips out of the fat with the spider, transfer to the colander, and shake to encourage any fat to drain off. Transfer to the paper towel–lined bowl, season generously with salt, and toss. Allow to drain for 30 seconds, then transfer to the serving bowl. Repeat with the remaining chips in three more batches, maintaining the oil temperature at 350°F (which is the tricky part).

4. Serve as soon as possible, but the chips should remain crisp for 1 or 2 days, depending on the weather.

CHEDDAR CHEESE CRACKERS

My son makes his own crackers, but he often has to take a break from making them because as he says, "Giving a young man the power to make his own salty snacks is dangerous." You've been warned: These are dangerously delicious. Bacon grease works well here, though it doesn't with all baked goods. The timing very much depends on how thinly the crackers have been rolled, so keep an eye on them during the final five minutes of baking.

Makes about 48 crackers

8 ounces sharp cheddar cheese, grated
¼ cup (1.7 ounces/50g) any animal fat
1 cup unbleached all-purpose flour
1 teaspoon baking powder
1 teaspoon fine sea salt
1 teaspoon garlic powder
1 teaspoon onion powder
2 tablespoons very cold water

1. Combine the cheese, fat, and ½ cup of the flour in a food processor and process until the mixture is well blended and looks like tiny pebbles.

2. Add the remaining ½ cup flour, the baking powder, salt, garlic powder, and onion powder, and begin processing. With the motor running, add the water and process until the dough comes together in a ball. Even if all the mixture doesn't form a ball, you should be able to gather it all together to form the ball.

3. Divide the dough in half and pat each half into a flattened rectangle, using a bench knife to keep the edges straight. Wrap in plastic wrap and chill for at least 30 minutes, or up to 1 day.

4. Preheat the oven to 375°F. Line a half sheet pan with parchment paper.

5. Because you don't want to add more flour, which would toughen the crackers, roll out the dough between two sheets of parchment paper or waxed paper if the dough is sticky, turning the dough over frequently. Roll each piece of dough as thinly as possible, ⅛ inch or less. A pasta roller works magic if you have one, but the dough is easy to handle and can be rolled out by hand readily. Use a pastry wheel, pizza cutter, or bench knife to cut the dough into 2-inch squares and transfer to the prepared sheet pan, leaving about ⅛ inch of space between them. Gather together the scraps, reroll, and cut into more crackers. Dock the crackers with a fork or toothpick.

6. Bake for 15 to 20 minutes, until puffed and very lightly browned. Sample a cracker. If it isn't as crisp as you want, return them to the oven for 2 to 3 minutes longer. Transfer the crackers to a cooling rack. Store in an airtight container.

WHIPPED LARDO

When you see "whipped lardo" on a restaurant menu, you might expect it to be made from lardo, a fatback cured with rosemary and other herbs and aged for at least six months (see page 105). It is not (usually). It is generally made from rendered lard or from unrendered fatback, seasoned with salt, a whisper of herbs, and a drop or two of high-quality vinegar. You can call it pork butter if you like. For such a simple recipe, however, there must be a few cautionary notes. First, if you are thinking of making this from unrendered fat, it must be fatback because that fat is free from the papery collagen ribbons that would ruin the texture of the whipped lardo. And it really should be from a Mangalitsa pig, which is uncommonly sweet with a softer texture of fat than the fat from other breeds. Second, to achieve the best texture, you must grind the fat, then whip it in a stand mixer — and that means a lot of cleanup. Also, it means that all of your utensils should be chilled in the freezer overnight. Finally, you might want to serve this piped out of a piping bag fitted with a large star tip and not tell your guests what you are serving them until after they've tasted it with an open mind; people tend to freak out about eating pure fat on bread, though they do it with butter all the time. Pickled vegetables are a good accompaniment.

Makes about 1 cup

About 10 ounces (284g) rendered fatback or leaf lard or unrendered fatback (preferably Mangalitsa fatback)

2 garlic cloves, minced

¼ teaspoon white wine vinegar, or as needed

1 teaspoon finely chopped fresh or dried rosemary

Maldon or other coarse sea salt and freshly ground black pepper

Sliced baguette or crackers, for serving

1. Grind the fat and garlic through the smallest setting of a well-chilled meat grinder or shred the fat with a well-chilled grating disc on a food processor. Add the vinegar and rosemary and beat with a wooden spoon. Cover and chill for at least 1 hour.

2. Put the fat in a stand mixer fitted with the whip attachment. Beat until creamy and light. Fold in salt and pepper to taste. This will need more salt than you might expect.

3. Pack or pipe the lardo into a ramekin, chill, and serve with a sliced baguette or crackers. Use within 1 week or put in an airtight container and freeze for long-term storage.

Until very recently, lardo was a food that was never found outside of Italy or Spain. When Mario Batali introduced it to his menu in his Manhattan restaurant Babbo, he instructed waiters to refer to lardo as "prosciutto bianco," aware that his customers would probably reject the idea of eating pure cured fat. Today lardo can be found in American restaurants and specialty food stores, but the ineffable quality of Italian lardo is sometimes lost in translation.

Lardo is a type of charcuterie made by curing strips of fatback with sea salt, black pepper, rosemary, garlic, and other herbs. The salt draws the moisture from the fat to create a brine that protects the fat from air and spoilage organisms. It can be aged for anywhere from 6 months to 2 years and will keep in the refrigerator for at least 6 months after it is cured, provided it is tightly wrapped to exclude air.

Purists insist that lardo can be made only from the fatback of a pig that has lived for at least 2 years and has never been fed corn. Only after a pig is 18 months old does it begin to develop fat that is tasty enough to be cured. The fat must be cured in the dark at cool temperatures. The most famous lardo, *lardo di Colonnata*, was made in marble boxes from the nearby Carrara marble mines. (You'll see recipes on the Internet that call for black plastic garbage bags to be used instead . . . not quite the same.) *Lardo di Colonnata* is now included in Slow Food's Ark of Taste catalog of heritage foods and has had IGP (Protected Geographical Indication) status since 2004.

Lardo is usually sold in slabs, to be sliced wafer-thin and served as part of an antipasti platter or as a topping for pizza or pasta. The texture is delicate and creamy while the flavor is mild, slightly sweet, and slightly herbaceous.

Lardo Ibérico de Bellota is the Spanish version of Italian lardo — same recipe but made from the fatback of Black pigs (a heritage pig breed) that have fed exclusively on acorns (*bellota*) for the last 5 months of their lives. The resulting lardo is softer than Italian lardo (almost liquid at room temperature) and sweetly flavored.

PARMESAN–ROSEMARY CRACKERS

Because Parmesan cheese is aged and dried longer than cheddar, it requires a slightly different recipe. Watch carefully during the final five minutes of baking and do not allow them to become too brown.

Makes about 48 crackers

4	ounces Parmesan or Romano cheese (or a combination of the two), grated
¼	cup (1.7 ounces/50g) any animal fat
1	cup unbleached all-purpose flour
1	teaspoon fine sea salt
1	teaspoon baking powder
1	teaspoon garlic powder
1	teaspoon onion powder
½	teaspoon fresh or dried rosemary, very finely chopped
2–2½	tablespoons very cold water

1. Combine the cheese, fat, and ½ cup of the flour in a food processor and process until well blended.

2. Add the remaining ½ cup flour, the salt, baking powder, garlic powder, onion powder, and rosemary, and begin processing. With the motor running, add 2 tablespoons of the water and process until the dough comes together in a ball. If the mixture does not form a ball (or a partial ball and will not hold together if pressed into a ball), add the remaining ½ tablespoon water. Even if all the mixture doesn't form a ball, you should be able to gather it all together by hand to form the ball.

3. Divide the dough in half and pat each half into a flattened rectangle, using a bench knife to keep the edges straight. Wrap in plastic wrap and chill for at least 30 minutes, or up to 1 day.

4. Preheat the oven to 375°F. Line a half sheet pan with parchment paper.

5. Because you don't want to add more flour, which would toughen the crackers, roll out between two sheets of parchment paper or waxed paper. Roll each piece of dough as thinly as possible, ⅛ inch or less, turning it over frequently. The dough is easy to handle and can be rolled out by hand readily. Use a pastry wheel, pizza cutter, or bench knife to trim the dough to make straight edges and cut the dough into 2-inch squares and transfer to the prepared sheet pan, leaving about ⅛ inch space between the crackers. Gather the scraps together and roll and cut until all the dough is used. Dock the crackers with a fork.

6. Bake for about 20 minutes, or until puffed and lightly browned. Transfer the crackers to a cooling rack. Store in an airtight container.

ONION CONFIT

Confit is a food slowly simmered in fat — the method is adapted from the famous duck confit of France. Cooking in any poultry fat gives these onions mellow flavor and a silken texture. Serve onion confit as a topping for steaks and burgers, omelets, and baked potatoes. It makes a delicious hors d'oeuvre on a toasted baguette. It can be used as the topping for pizza; add olives and anchovies and you have the classic topping for pissaladière.

Makes 2 cups

- 1½ pounds pearl or boiling onions
- ½ cup (3.5 ounces/100g) any poultry fat
- 3 thyme sprigs, or 1 teaspoon dried thyme
- 1 tablespoon brown sugar
- ½ teaspoon fine sea salt, plus more as needed
  Freshly ground black pepper

1. To peel the onions, put the onions in a large heatproof bowl and pour boiling water over them. When the onions are cool enough to handle by hand, the skins should slip off easily. Slice off the root ends and cut larger boiling onions into halves or quarters to make bite-size pieces.

2. Preheat the oven to 275°F.

3. Melt the fat in a Dutch oven over medium-high heat. Add the onions, thyme, sugar, salt, and several grinds of pepper. Stir to coat the onions in the fat. Cover and transfer to the oven.

4. Bake for 2 hours, until the onions are completely tender. Remove the lid and bake for 30 minutes longer, until the onions begin to brown. Discard the thyme sprigs if used.

5. Transfer the onions to a crock or pint canning jar for storage. Serve warm or at room temperature.

JEWISH-STYLE CHOPPED LIVER

When I was growing up, no holiday was complete without chopped liver, usually served as a round scoop placed on a leaf of lettuce and accompanied by challah or matzoh. These days I am more likely to serve it in a ramekin, surrounded by crackers or slices of baguette. If calling it liver pâté makes it taste better to you, go right ahead.

Makes about 3 cups

- 1¼ pounds chicken, goose, or duck livers
- ¼ cup (1.7 ounces/50g) any poultry fat
- 2 large onions, coarsely chopped
- 1 tablespoon brandy, sherry, or cognac
- 2 eggs, hard-cooked and peeled
- 1 teaspoon fine sea salt, plus more as needed
  Freshly ground black pepper
  Crackers or sliced baguette, for serving

Recipe continues on next page

1. Wash the livers under cold running water to remove all traces of blood. Pat dry. Trim the livers by removing any greenish or blackish spots and any membranes. Cut into 1-inch pieces.

2. Melt the fat in a large skillet over medium heat. Add the livers and sauté until the livers are browned on the outside, but still rosy inside, 3 to 5 minutes. Do not sear the livers; the skins should remain soft. With a slotted spoon, transfer the livers to a food processor.

3. Add the onions to the fat remaining in the skillet and cook until lightly browned, 6 to 8 minutes. Add to the livers in the food processor.

4. Add the brandy to the food processor and pulse the mixture about a dozen times to make a coarse paste. Transfer to a bowl. Put the eggs in a food processor and pulse to finely chop, then mix into the liver. Season generously with the salt and pepper.

5. Cover with plastic wrap and chill before serving. Serve with crackers or sliced baguette.

   NOTE: Because liver is very perishable, do not make more than you will consume

in 3 or 4 days. If you want, you can freeze extra. The frozen and then thawed chopped liver may need moistening, which you can add in the form of melted poultry fat, a little olive oil, or even a little chicken broth.

PÂTÉ OF GIZZARD CONFIT

Here's a pâté that doesn't require pork and instead calls for chicken gizzards — a completely underutilized part of the chicken. As with all pâtés, serve with good bread and pickles to cut the richness.

Serves 8–10

> Confit of Gizzards with its fat (page 260)
> ¼ cup chopped fresh parsley
> ½ cup gribenes (poultry cracklings), for serving (optional)
> French bread, sliced, for serving
> Pickles, for serving
> Whole-grain mustard, for serving

1. Finely chop the gizzards and place in a small bowl. Stir in the parsley and enough reserved fat to make a spreadable mixture. If desired, crisp the gribenes in a skillet over medium heat for about 5 minutes; do not burn.

2. Pack the pâté into a small crock or bowl and top with the crisped gribenes. Serve with bread, pickles, and mustard on the side.

Early American literature is filled with depictions of poor and working people eating toast and drippings. American writers wrote about bread and drippings to convey the poverty of the diet — bread with the leftover juices of cooked meat or bacon. But don't knock it till you've tried it. We aren't talking about rendered fat that is kept as flavor-free as possible. We are talking about the fat that is left after a luxurious roast of any meat or poultry.

In more frugal times, people kept a "drippings pot" in the fridge (or in a cool larder or pantry). They used the collected fats for all sorts of cooking. The drippings made from a well-seasoned roast have tons of flavor! Here are several combinations you would be proud to serve as an appetizer or hors d'oeuvre at a dinner party — or to make a simple supper accompanied by salad or cooked veggies.

Start by preheating the oven to 350°F. Thinly slice a baguette on the diagonal and arrange the slices on a sheet pan in a single layer. Drizzle both sides with a little melted fat (preferably whatever type of drippings you are using). You don't need to saturate the bread or provide complete coverage — a light drizzle should help with the browning. Season both sides with salt and pepper. Toast the bread, rotating the sheet pan from front to back and flipping the slices halfway through baking, for about 12 minutes, until the toasts are browned and crispy throughout.

POULTRY FAT AND GRIBENES: Spread the toast with chicken, duck, or goose fat and sprinkle on cracklings. Top with pickled cabbage or another pickle to cut the richness.

POULTRY FAT AND APPLE SPREAD: Peel and dice a small apple and peel and dice a small onion. Sauté in any poultry fat over medium heat until the onion caramelizes and the apple softens completely. Chill until firm, then blend or whip until you have the consistency of whipped butter.

BEEF OR PORK DRIPPINGS: After roasting any cut of beef or pork, remove the meat from the roasting pan. Transfer the liquid that remains into a fat separator. Allow the fat to rise to the top and separate from the meat juices. Scrape off the fat and beat the fat with a fork until it has an even texture.

LARD SPREADS: *Smalec* is a Polish spread made from lard sautéed with apples, onion, and garlic. In southern Spain, you can find *manteca colorá* (Andalusian for "red lard"), a spread made from lard and finely chopped pieces of pork flavored with herbs, usually bay leaf and oregano. Paprika gives the dish its characteristic orange color and flavor. *Griebenschmalz*, lard with crispy cracklings, is a German spread. Whipped Lardo (page 104) is whipped lard with herbs. Lardo, cured lard with herbs (see page 105), is more likely found very thinly sliced and served on toast.

PORK RILLETTES

The French Renaissance writer Rabelais called rillettes "the brown jam from the pig" (*brune confiture de cochon*), and that's as good a definition as any for this spread made by simmering pork shoulder (or belly) in its own fat, then chopping until it has a spreadable texture. After it is made, it is packed into a crock and sealed with a generous layer of fat. By using the fat to exclude oxygen, a well-packed jar of rillettes will keep in the refrigerator for about a month, ready to be served on toasted or fresh bread as an appetizer alongside cornichons, whole-grain mustard, and fruit preserves or mostarda.

Makes 4 cups

- 6 garlic cloves, minced
- 1 tablespoon fine sea salt
- ½ teaspoon freshly ground black pepper
- ½ teaspoon freshly ground white pepper
- ½ teaspoon ground ginger
- ¼ teaspoon freshly grated nutmeg
- ⅛ teaspoon ground cloves
- ¼ cup dry red or white wine
- 3 pounds boneless pork shoulder, cut into 1-inch cubes
- 1½ cups (10.5 ounces/300g) lard
- 4 bay leaves
- 6 thyme sprigs
  About ½ cup pork or chicken broth

1. Combine the garlic, salt, black and white peppers, ginger, nutmeg, cloves, and wine in a bowl. Add the pork cubes and massage the spices into the meat until it is well distributed.

2. Preheat the oven to 275°F.

3. Melt the lard in a Dutch oven over medium heat. Add the meat with its spices. Tuck the bay leaves and thyme among the meat pieces. If the lard does not cover the meat completely, add enough broth to cover. Cover the pot, place in the oven, and cook for 3 hours, until the pork is completely tender; test it after 2½ hours. The meat should be succulent and falling-apart tender, but not dry.

4. Remove the pan from the oven and discard the bay leaves and thyme. Set a large strainer over a heatproof bowl and carefully pour the pork mixture into it, reserving the drained fat and juices.

5. Transfer the pork chunks to the bowl of a stand mixer fitted with the paddle attachment. Turn the mixer on to low speed and gradually increase the speed to medium, allowing the pork to break down and shred. Slowly drizzle in about one-third of the reserved fat and juices. If the mixture is not as loose and creamy as you would like it, slowly add more of the liquid, a tablespoon at a time, until it reaches the desired texture.

Recipe continues on next page

6. Pour the remaining liquid into a fat separator and allow to stand until the pork juices have separated from the lard.

7. Carefully pack the mixture into jars or crocks, spooning it in a little bit at a time and making sure to remove all air bubbles. Smooth the tops of the mixture with the back of a spoon, wipe the rims of the jars with a clean cloth, then pour at least ¼ inch of fat on top of each one. Close the lids and refrigerate for at least 2 hours before serving, or up to 1 month. Any remaining lard can be reserved for stovetop cooking and stored in the refrigerator for up to 1 week or frozen for longer keeping. Any remaining pork juices can be reserved to enrich a soup or sauce and reserved for up to 5 days in the refrigerator or frozen for longer keeping. The rillettes can also be frozen directly in their jars and stored for several months. Thaw in the refrigerator overnight before serving.

BACON JAM

Bacon jam is a relish that is at home on top of a burger, on a crostini, or on a cracker. It is also pretty terrific on a toasted peanut butter sandwich. There are many versions of this recipe out there, and most are slow-cooked. My version is streamlined, with the secret ingredient of soy sauce giving it a hit of umami and joining the flavors together.

Makes about 1¼ cups

- 8 ounces bacon, diced
- 1 onion, thinly sliced
- 2 garlic cloves, minced
- ½ cup pure maple syrup
- ¼ cup balsamic vinegar
- 2 tablespoons apple cider vinegar
- 1 teaspoon soy sauce, or to taste
- ½ cup beef or chicken broth

1. Brown the bacon in a large heavy saucepan or Dutch oven over medium heat until the fat has rendered, about 8 minutes.

2. Add the onion and sauté until golden, about 6 minutes. Add the garlic, maple syrup, balsamic vinegar, cider vinegar, and soy sauce, and stir to combine. Add the broth and bring to a simmer, scraping up the browned bits from the bottom of the pot. Continue to simmer, stirring occasionally, until the mixture has thickened, about 10 minutes.

3. Transfer the mixture to a blender or food processor and pulse until well combined, leaving some texture. Return the mixture to the saucepan, place over medium-high heat, and cook, stirring occasionally, until it has reduced to a spreadable, jamlike consistency, about 5 minutes. Remove from the heat, taste, and add more soy sauce if needed.

Cool and serve or store in the refrigerator in an airtight container for up to 2 weeks. Let warm to room temperature before using.

SCALLION PANCAKES

In China, scallion pancakes are a breakfast food, often found as a grab-and-go item in a food court. In this country, the pancakes are more often found among the appetizers at a Chinese restaurant. No reason you can't serve them for breakfast or for a snack. Don't think of them as pancakes, however; think of them as flaky, fried flatbreads.

Serves 4–8

DOUGH

- 2½ cups unbleached all-purpose flour
- 1 teaspoon fine sea salt
- 3 tablespoons (1.4 ounces/40g) plus ¼ cup (1.7 ounces/50g) lard
- 1 cup lukewarm (110°F) water

DIPPING SAUCE

- ¼ cup soy sauce
- ¼ cup Chinese black vinegar
- 1 teaspoon peeled and minced fresh gingerroot
- 1 teaspoon Chinese chili-garlic sauce
- 2 teaspoons sugar

- 1 tablespoon toasted sesame oil
- 1 bunch (6–8) scallions, white and green parts, finely chopped

1. To make the dough, pulse the flour and salt in a food processor to combine. Add the 3 tablespoons lard and pulse to mix in. With the machine running, slowly drizzle in the water until the dough comes together.

2. Transfer the dough to a floured work surface and knead to form a soft, smooth ball. Transfer to a greased bowl, cover with a damp towel, and allow to rest for 30 minutes at room temperature.

3. While the dough rests, preheat the oven to 200°F. Place a wire rack on a sheet pan.

4. To make the sauce, combine the soy sauce, vinegar, ginger, chili-garlic sauce, and sugar in a small bowl and set aside at room temperature.

5. On a lightly floured work surface, roll out the dough as thinly as possible to form a large square. Brush with the sesame oil and sprinkle with the scallions. Fold in half and press gently to flatten. Fold in the sides to meet in the middle and press gently to flatten. Then fold one side over the other to form a rectangle. Fold the bottom of the rectangle up to the top to form a square with multiple layers. Divide the dough into eight pieces and gently form each piece into a smooth ball. Flatten the pieces by hand to form disks.

Recipe continues on page 115

Scallion Pancakes, continued

6. Melt the remaining ¼ cup lard in an 8-inch cast-iron skillet over medium-high heat until shimmering. Working with one dough disk at a time, press, roll, or stretch the dough into a pancake 4 to 5 inches in diameter and slip the pancake into the hot fat. Cook until you can see the edges beginning to brown, about 2 minutes. Carefully flip with a spatula or tongs (be careful not to splash the fat) and continue to cook until the second side is an even golden brown, 1 to 2 minutes longer. Transfer to the wire rack and keep warm in the oven. Repeat to fry the remaining pancakes.

7. To serve, cut each pancake into wedges. Serve hot, passing the dipping sauce at the table.

POTATO KNISHES

Like the Cornish pasty, a knish has a sturdy crust and may be filled with meat, vegetables, cheese, or potatoes, as it is here. Some knishes, unlike pasties, may be filled with kasha (buckwheat groats), which reveals its Jewish Eastern European origins. Though originally a workingman's lunch, these days knishes are usually *noshes* (snacks) rather than main dishes. In America, knishes became popular as small treats sold from pushcarts in New York City in the late 1800s, the original New York food carts. From the pushcarts, knishes traveled to Jewish delis, where mass-produced versions can still be found.

Makes 8 knishes

DOUGH

2½ cups unbleached all-purpose flour
1 teaspoon fine sea salt
1 egg, beaten
½ cup (3.5 ounces/100g) any poultry fat
½ cup water
1 teaspoon distilled white vinegar

FILLING

1½ pounds (3–4 medium) russet potatoes, peeled and cut into chunks
Fine sea salt
¼ cup (1.7 ounces/50g) any poultry fat
2 onions, peeled and diced
Freshly ground black pepper

1 egg, beaten with 1 tablespoon water, for egg wash

1. To make the dough, whisk together the flour and salt in a large bowl. Make a well in the center and pour in the egg, fat, water, and vinegar. Stir to combine, pushing more and more flour into the center until all the flour is moistened. Gather the dough into a ball and knead until smooth. Return the

Recipe continues on next page

dough to the bowl and cover with plastic wrap. Set it aside while you prepare the filling.

2. To make the filling, put the potatoes in a large saucepan, cover with cold water, add salt generously, and bring to a boil over high heat. Reduce the heat to medium and cook until the potatoes can be pierced easily with a knife, about 20 minutes. Drain and return to the saucepan.

3. While the potatoes cook, heat a medium skillet over medium heat. Add 3 tablespoons of the fat and melt. Add the onions and cook until golden, 10 minutes. Remove from the heat.

4. Mash the potatoes with a potato masher or run them through a ricer. Scrape in the onions and fat from the skillet. Mix well and season with salt and pepper to taste.

5. Preheat the oven to 400°F. Line a sheet pan with parchment paper.

6. To assemble the knishes, divide the dough in half and flatten each half into a flattened disk. On a very lightly floured surface, roll and stretch one piece of the dough into a very thin sheet, roughly a 1-foot square. Spoon a 2-inch-wide line of filling across the bottom of the dough, leaving a border of about 1½ inches. Pick up the dough from the border and fold it over the filling. Brush some of the remaining 1 tablespoon fat over the dough. Tightly roll the filling toward the top edge of the dough, brushing with fat on the top as you roll. Keep rolling until the filling has been wrapped at least twice in dough. Trim the ends of the dough so that they're even with the potato filling. Mark indentations on the log every 3 inches and use a sharp knife or bench knife to cut the roll into four pieces. Pick up one piece and pinch together the dough to enclose one of the ends. Use the palm of your hand to flatten the knish a bit into a squat, almost square shape. Pinch together the other open end to seal. Place on the prepared sheet pan. Repeat with the remaining pieces, then repeat the process with the remaining piece of dough, filling, and fat.

7. Brush the egg wash over the knishes. Bake for 40 to 45 minutes, until golden brown. Serve warm.

TOMATO AND MOZZARELLA PANZAROTTI

Panzarotti translates from the Italian as "little bellies," which doesn't begin to describe these wonderful pizza puffs. Filled with a savory filling of tomato sauce and mozzarella, these deep-fried hot pockets are meant to be eaten hot from a food truck but hold up surprisingly well when cool. If I've said it once, I've said it a thousand times: Foods fried in lard just aren't greasy, which makes these perfect picnic foods, after-school snacks, and midnight delights.

Makes 12 panzarotti

DOUGH

3¼ cups unbleached all-purpose flour
1 tablespoon fine sea salt
1 tablespoon instant or quick-rise yeast
3 tablespoons (1.4 ounces/40g) lard
1 cup lukewarm (100°F) water

FILLING

3 tablespoons (1.4 ounces/40g) lard
½ onion, diced
½ green bell pepper, diced
4 garlic cloves, minced
½ teaspoon red pepper flakes (optional)
½ cup seasoned pasta sauce
Leaves from 1 basil sprig, chopped
10 Kalamata olives, pitted and chopped
4 oil-packed anchovy fillets, chopped
1 tablespoon capers
1¼ cups shredded mozzarella cheese
2 cups (14 ounces/400g) lard or tallow

Recipe continues on next page

SALO

Americans' love affair with Italian foods has meant that many are familiar with lardo, but salo, a similar specialty of Eastern Europe, is much harder to find — or even to learn about.

Salo (as it is known in the Ukraine and in Russia) is cured slabs of fatback (rarely pork belly), with or without skin. Salo, known by other names, is fairly common across Eastern Europe, with some variations in seasonings; it is also sometimes smoked.

Whereas lardo is generally seasoned with rosemary and served raw in paper-thin slices, salo may be seasoned with paprika, garlic, black pepper, or coriander and may be served raw or cooked. It is often fried with finely chopped garlic as a condiment for borscht (beet soup). Sometimes it is fried to render out the fat, with the remaining cracklings served atop fried potatoes or spread on bread. Thinly sliced salo on rye bread rubbed with garlic is a traditional snack to accompany vodka in Russia and *horilka* (a moonshine type of vodka) in Ukraine.

1. To make the dough, mix the flour with the salt and yeast in a food processor fitted with the dough blade. Add the lard and process until evenly combined. With the machine running, add the water and process until the dough comes together in a ball, scraping down the sides of the bowl as needed. Transfer to a lightly floured surface and knead a few times, until smooth. Place in an oiled bowl, cover with a kitchen towel, and leave for about 2 hours, until the dough has doubled in volume.

2. Line a half sheet pan with parchment paper. Divide the dough into 12 pieces and then shape each piece into a ball. Place the dough balls on the prepared sheet pan, cover with a towel, and let stand for 1 hour.

3. Meanwhile, make the filling. Melt the lard in a small saucepan over medium heat. Add the onion, green pepper, garlic, and pepper flakes (if desired), and cook, stirring, until softened, about 2 minutes. Add the pasta sauce, basil, olives, anchovies, and capers, and cook, stirring, for about 5 minutes, to blend the flavors.

4. On a clean work surface, roll each dough ball into a 5-inch disk. (If the dough sticks, lightly oil the work surface and rolling pin, but you probably won't need it.) Place 1 heaping tablespoon of the filling on one half of each disk, then top with a heaping tablespoon of the cheese. Fold the dough over the filling and pinch the edges to seal. Roll the edges over and twist to form a seal that resembles a braid.

5. Line a half sheet pan with wire racks. Melt the lard in a large skillet and attach a deep-fry thermometer to the side of the pan; heat the lard to 350°F. Working with a few panzarotti at a time, add to the lard and fry, turning once, until golden brown on both sides, 4 to 5 minutes. Remove the panzarotti from the pan and allow to drain on the wire racks. Serve hot.

CHORIZO–CHEESE EMPANADAS

Tasty handheld hot pockets, empanadas are found throughout the world, thanks to the intrepid Spanish explorers. A popular street food, empanadas are made throughout Central and South America and the Philippines. If you take some on a picnic, they will disappear faster than you might imagine. I love having them on hand for a quick lunch, or even a grab-and-go breakfast.

Makes about 20 empanadas

   1 tablespoon (0.45 ounce/13g) lard or tallow
14–16 ounces fresh chorizo, removed from its casings

1 onion, finely chopped

1 red or green bell pepper, seeded and finely chopped

1 chile, such as Hungarian hot wax, jalapeño, serrano, or Fresno, seeded and finely chopped

¼ cup salsa, homemade or store-bought

2 teaspoons dried oregano

12 ounces shredded Monterey Jack cheese
    Salt and freshly ground black pepper
    Pasty Pastry (page 273)

1 egg, beaten with 1 tablespoon water, for egg wash

1. Melt the fat in a large skillet over medium-high heat. Add the chorizo and cook, breaking it up with a spoon, until mostly browned, about 5 minutes. Add the onion, bell pepper, and chile to the skillet and sauté until tender, about 5 minutes. Transfer to a bowl and stir in the salsa and oregano. Cover and chill the filling until ready to use.

2. Preheat the oven to 400°F, using the convection function if you have one. Line two half sheet pans with parchment paper. Remove the chorizo mixture from the refrigerator and mix in the cheese. Taste and add salt and pepper as needed.

3. Working with one log of dough at a time, evenly divide the log into 10 pieces. On a lightly floured surface, roll out each piece of dough into a round with a thickness of about ⅛ inch. Using a 4½-inch saucer or template, cut out each round with a knife. Holding the dough round in your hand, brush the outer rim of each dough circle with the egg wash. Spoon a heaping tablespoon of filling into the center. Pull up the sides of the dough circles to meet on top of the filling. Pinch the edges together to form a crimped and fluted seal. Set the empanada, crimped seal up, on the prepared sheet pan. Allow about 1 inch between the empanadas. Although you may be able to squeeze all the empanadas onto one half sheet pan, the empanadas will bake more evenly if given more room on two pans. Gather together the dough scraps and reroll, making more empanadas until all the filling is used.

4. When all the empanadas are formed, cut slits into the sides to allow steam to escape during baking. Brush the remaining egg wash over the empanadas.

5. Bake in the middle of the oven for 15 minutes. Decrease the oven temperature to 350°F and bake for 30 to 35 minutes longer, until the empanadas are golden, rotating the sheets about halfway through the baking.

6. Don't be alarmed if the empanadas have released fats and juices. Transfer the empanadas to a paper towel–lined sheet pan and let cool for at least 10 minutes before serving.

ARGENTINEAN BEEF EMPANADAS

The variations on empanadas found throughout Central and South America are endless. In Argentina, they are typically made with beef, and often the filling also contains olives and raisins. The crust is traditionally made with beef fat, which gives the empanadas a deliciously crisp and flaky crunch. They make a great lunch or light dinner with a salad, but they are most often found in South America as a street food, meant to be enjoyed as a hearty snack. They freeze well, baked or unbaked. If you have frozen some, bake them still frozen, but add to the baking time. In my house, they disappear very, very quickly.

Makes about 20 empanadas

- 1 tablespoon (0.45 ounce/13g) any animal fat
- 1 pound ground beef
- 1 onion, finely chopped
- 1 red or green bell pepper, seeded and finely chopped
- 1 mild chile, such as Hungarian hot wax, seeded and finely chopped
- 1 tablespoon ground cumin
- 1 tablespoon sweet paprika
- 1 tablespoon dried oregano
- 1 cup beef or chicken broth
- 1 tablespoon sugar
- ½ cup raisins
- ½ cup chopped green olives with pimento
- 1 egg, hard-cooked, peeled, and finely chopped
- Salt and freshly ground black pepper
- Pasty Pastry (page 273)
- 1 egg beaten with 1 tablespoon water, for egg wash

1. Melt the fat in a large skillet over medium-high heat. Add the beef and cook, breaking it up with a spoon, until browned but not completely cooked through, about 5 minutes. Add the onion, bell pepper, and chile to the skillet and sauté until tender, about 5 minutes. Add the cumin, paprika, and oregano, and sauté until fragrant, about 1 minute. Add the broth and bring to a boil, stirring and scraping up any browned bits. Continue to boil, stirring occasionally, until most of the liquid is evaporated, 15 to 20 minutes. Stir in the sugar, raisins, olives, and hard-cooked egg, and season with salt and pepper. Transfer to a bowl, cover, and chill the filling until ready to use.

2. Preheat the oven to 400°F, using the convection function if you have one. Line two half sheet pans with parchment paper.

3. Working with one log of dough at a time, evenly divide the log into 10 pieces. On a lightly floured surface, roll out each piece of dough into a round with a thickness of about ⅛ inch. Using a 4½-inch saucer or template, cut out each round with a knife. Holding the dough round in your hand,

brush the outer rim of each dough circle with the egg wash. Spoon a generous tablespoon of filling into the center. Pull up the sides of the dough circles to meet on top of the filling. Pinch the edges together to form a crimped and fluted seal. Set the empanada, crimped seal up, on the prepared sheet pan. Allow about 1 inch between the empanadas. Although you can probably squeeze all the empanadas onto one half sheet pan, the empanadas will bake more evenly if given more room on two pans. Gather together the dough scraps and reroll, making more empanadas until all the filling is used.

4. When all the empanadas are formed, cut slits into the sides to allow steam to escape during baking. Brush the remaining egg wash over the empanadas.

Recipe continues on next page

HOLY HOT POCKETS

While hot pockets are a joke, a tired meme, and Hot Pockets are a successful trademarked brand of mass-produced, microwavable savory pastries in the United States, made-from-scratch hand pies — baked or fried — are enjoyed the world over. Often sold on the street or at bakeries for a grab-and-go snack or lunch, such portable pies are a wonderful way to sample a foreign cuisine, at home in your own kitchen or when traveling.

Hot pockets come in all sorts of shapes and sizes, but all descend from the original large pies, developed in the early days of baking in ovens, when the pastry wrap was meant to protect the meats and vegetables in the filling from the erratic heat of a wood-fired oven. While family-size pies, sweet and savory, remain popular, a subset of hand-size pies go by the names of pasties and empanadas (and others) and provided working folks with a lunch that was easily transported.

Although modern recipes often call for oil, shortening, butter, or margarine(!) in the crust, not to mention commercial piecrust, puff pastry sheets, and empanada disks, most originated using flour and the convenient animal fats found in the home. For Cornish pasties, that meant a beef suet (tallow) crust, as it did for Argentinean empanadas. For potato knishes, that meant using chicken schmaltz or the commonly available goose fat in the Old World. Lard was typically used in many hand pies, included the fried hand pies of the American South. Eggs are frequently added to the pastry dough to give it strength, and the pastry may be deep-fried or baked.

Argentinean Beef Empanadas, continued

5. Bake in the middle of the oven for 15 minutes. Decrease the oven temperature to 350°F and bake for 30 to 35 minutes longer, until the empanadas are golden, rotating the sheets about halfway through the baking.

6. Don't be alarmed if the empanadas have released fats and juices. Transfer the empanadas to a paper towel–lined sheet pan and let cool for at least 10 minutes before serving.

EDIBLE CANDLES

You have probably enjoyed bread dipped in olive oil at high-end (and not-so-high-end) Italian restaurants. A new twist on bread dipping is the edible beef tallow candle, sometimes plain, sometimes blended or infused with herbs or garlic. The candle is often a short one, made in a mold like a shot glass (or sometimes a hollowed-out marrow bone) and set on a small saucer. The waiter lights the candle and brings the bread basket to your table. As the candle melts, you are invited to dip your bread into the pool of delicious melted fat. This is nose-to-tail eating at its finest.

KOREAN FRIED CHICKEN WINGS

A bar snack par excellence, the tender wings are coated in a sauce that is a balance of sweet, hot, sour, and salty. The hot comes from the distinctively flavored Korean hot pepper paste, which is the same pepper paste that gives many kimchis their distinctive flavors. It is available wherever Asian foods are sold and is a great seasoning to have on hand. Serve the wings with plenty of beer and napkins. *See photo, page 96.*

Serves 4

- 4 pounds chicken wings
  Salt and freshly ground black pepper
- ½ cup cornstarch
- ¼ cup Korean gochujang (red chili paste)
- ¼ cup honey
- ¼ cup rice vinegar
- 2 tablespoons soy sauce
- 2 tablespoons hot water
  Lard or tallow, for deep-frying
  White sesame seeds, for garnish

1. Separate the chicken wings into wings and drumettes. Reserve the bony tips for broth. Place the chicken in a large bowl, season with salt and pepper, add the cornstarch, and toss to coat. Set aside.

2. To prepare the sauce, combine the gochujang, honey, vinegar, soy sauce, and water in a large bowl and mix well. Set aside.

3. In a large deep Dutch oven or wok, melt enough fat to fill the pot by several inches and heat to 375°F. Set up a sheet pan lined with wire racks to drain the fried wings.

4. Preheat the oven to 200°F.

5. Toss the wings in the cornstarch again, then shake off any excess. Working in batches, slide the wings into the oil and fry until golden and crispy, about 8 minutes, stirring and turning the wings on occasion. Remove the cooked wings from the fat, place on the wire racks, and put the sheet pan in the oven to keep warm. Bring the fat back up to 375°F before adding the next batch. Continue until all the wings are cooked.

6. Just before serving, add the wings to the sauce and toss to coat well. Transfer the wings to a platter, garnish with sesame seeds, and serve immediately.

Summer Vegetable Galette,
page 126

MAIN DISHES

SUMMER VEGETABLE GALETTE

Galette or crostata? They are one and the same: a rustic tart with an exposed filling. Crusts made with lard are slightly easier to handle than crusts made with tallow, but either works here. Try to keep the edges even by rolling from the center out and turning the dough regularly.

The flaky pastry combined with the tasty pesto and roasted summer vegetables make this a showstopper. Serve this as a light summer meal with a green salad for lunch or dinner. *See photo, page 124.*

Serves 6

Double-Crust Pastry made with lard or tallow (pages 270 and 272)

2 tomatoes, thinly sliced

Salt

¼ cup (1.7 ounces/50g) any animal fat

1 small eggplant (about 1 pound), diced

1 zucchini (about 8 ounces), diced

1 onion, halved vertically and thinly slivered

Freshly ground black pepper

½ cup prepared pesto

About 12 pitted Kalamata olives

Small basil leaves, for garnish

1. Chill the pastry for at least 30 minutes.

2. Preheat the oven to 425°F. Place a half sheet pan in the oven to preheat at the same time.

3. To make the filling, put a layer of paper towels on a sheet pan and arrange the tomato slices on it in a single layer. Sprinkle with salt and set aside to drain.

4. Remove the hot sheet pan from the oven. Add the fat and tip and turn the pan until the fat is melted. Spread the eggplant, zucchini, and onion on the sheet pan and toss with two silicone spatulas until well coated with the fat. Spread out in a single layer and season lightly with salt and pepper. Roast for about 30 minutes, turning the vegetables occasionally, until the vegetables are tender and beginning to brown.

5. Reduce the oven temperature to 400°F.

6. Roll out the dough on a very lightly floured sheet of parchment paper to a 12-inch round; the edges can be ragged. Transfer the parchment paper and dough onto a half sheet pan. It is okay if the dough overlaps the edge of the pan.

7. Spread the pesto in a thin layer over the pastry, leaving a 1½-inch border all around. Spoon the vegetables evenly over the pesto. Scatter the olives over. Pat the sliced tomatoes with paper towels to blot any moisture and arrange on top. Fold up the edges of the crust, pleating as needed; the galette will be partly open, exposing the filling.

8. Bake about 40 minutes, until golden brown. Cool on a wire rack for at least 20 minutes. Just before slicing and serving, sprinkle with a small handful of basil leaves.

ANIMAL FATS SEASON CAST IRON PERFECTLY

Instructions for seasoning cast iron have you use any old cooking oil. In fact, vegetable seed oils leave a gummy residue when used for seasoning (or cooking in) cast iron. Lard or tallow is a better choice, giving a smooth, nonstick, not-gummy finish.

To season cast iron, first wash well and scrub away any rust with fine steel wool. Dry completely over low heat on top of the stove. Then rub the pan all over with melted tallow or room-temperature lard and then carefully rub the fat off (some will remain). Place in a cold oven and heat to 450°F. Leave for 30 minutes. Remove from the oven (wearing oven gloves!), carefully rub with fat again, then rub off the fat and bake for 30 minutes longer. Your pan is now seasoned and should not need seasoning again, as long as it is in active use and not allowed to rust.

ROASTED ROOT VEGETABLE TART

When fall rolls around, root vegetables are the way to go for vegetable pies — and roasting is the way to go with root vegetables. If you prep the vegetables while the dough chills, the timing works out perfectly. Do cut the vegetables into a small dice (no bigger than ½ inch) for even cooking and best appearance.

Serves 4–6

- ½ recipe for double piecrust made with any animal fat (except bacon grease, pages 270–72)
- 3 tablespoons (1.4 ounces/40g) any animal fat (poultry fat is recommended)
- 6 cups diced mixed roasted root vegetables (golden beets, carrots, celery root, parsnips, rutabagas, salsify, and/or turnips)
- 4 shallots, peeled and quartered
  Salt and freshly ground black pepper
- 3 eggs
- ¾ cup heavy cream or half-and-half
- ½ cup freshly grated Parmesan cheese

1. Keep the pastry dough chilled while you prepare the vegetables.

2. Preheat the oven to 450°F. Place a half sheet pan in the middle of the oven to preheat at the same time.

3. Add the fat to the hot sheet pan and tilt the pan until the fat is melted and covers the bottom of the pan. Add the vegetables and shallots and toss with two silicone spatulas until evenly coated with fat. Spread out in a single layer and season with salt and pepper. Roast for about 30 minutes, until tender and lightly browned, turning the vegetables once after about 20 minutes.

4. Roll out the pastry on a floured work surface until it is about ⅛ inch thick. Transfer to a pie plate and trim and crimp the edges. Put the pie shell in the refrigerator to keep the pastry chilled.

5. Beat the eggs with the cream in a small bowl.

6. When the vegetables are nicely browned, transfer into the chilled pie shell. Pour the egg mixture evenly over the vegetables. Top with the Parmesan cheese.

7. Put the pie in the oven and immediately reduce the oven temperature to 350°F. Bake for 35 to 40 minutes, until the filling is set and the top is lightly browned.

8. Let the tart set for at least 10 minutes before serving. Serve warm or at room temperature.

FRISÉE AND POTATO SALAD WITH CHICKEN GIZZARD CONFIT

Once you've made the chicken gizzard confit (or any confit), the question becomes how to serve it. This is a traditional French salad, usually made with duck or geese gizzards, which you can certainly use instead of the chicken gizzards. It can also be made with duck confit (page 261). The bitter greens are necessary to cut through the richness of the confit. You will use only 6 tablespoons of the chicken fat from the confit. Reserve the rest to use in sautés or pilafs, or in any egg or potato dish.

Serves 4

- 1½ pounds waxy potatoes, sliced about ¼ inch thick
- Coarse sea salt
- Confit of Gizzards (page 260)
- ¼ cup red wine vinegar
- 1 shallot, minced
- 1 teaspoon Dijon mustard
- Freshly ground black pepper
- 1 large head frisée or radicchio, trimmed and torn into pieces
- 2 tablespoons gribenes (poultry cracklings), for garnish (optional)

1. Put the potatoes in a saucepan, cover with cold water, and add salt generously. Bring to a boil over high heat. Reduce the heat and simmer until the potatoes are just fork-tender, 10 to 15 minutes; do not overcook or the slices will fall apart. Drain.

2. While the potatoes cook, warm the confit in a small saucepan until the fat is liquid, then remove from the heat.

3. Whisk together the vinegar, shallot, and mustard in a bowl. Pour in 6 tablespoons of the heated confit fat in a slow, steady stream, whisking constantly until emulsified. Season with salt and pepper. Add the potatoes and toss to coat. Add the frisée and the gizzard confit and toss to mix.

4. Transfer to a serving platter or individual plates. Sprinkle the gribenes over the top, if desired, and serve at once.

MATZOH BALL SOUP

Matzoh balls are dumplings for soup, made with matzoh meal (ground-up matzohs) instead of flour. It's a Jewish specialty, traditionally made with chicken fat (or goose fat in the Old World). Unfortunately, in recent years, the chicken fat in most recipes was replaced with a vegetable oil, resulting in a loss of texture and flavor. When I was growing up, the soup — broth plus matzoh balls — was served as a first course. But I prefer to add the chicken and vegetables for a one-dish meal that isn't reserved for holidays.

Serves 4–6

- ¼ cup (1.7 ounces/50g) chicken fat (or any poultry fat)
- ¼ cup water
- 4 eggs
- 1 teaspoon fine sea salt
- 1 cup matzoh meal
- 6–8 cups chicken broth
- 4 cups shredded greens (such as bok choy, chard, Chinese cabbage, escarole, kale, or spinach)
- 2 cups diced cooked chicken

1. Whisk together the fat, water, eggs, and salt in a medium bowl. Stir in the matzoh meal until well blended. Cover and refrigerate for at least 15 minutes, or up to 6 hours.

2. Meanwhile, bring a large pot of generously salted water to a boil. Reduce the heat to medium-low to keep the water gently boiling.

3. Form the chilled matzoh batter into 1-inch balls and carefully ease into the water. Handling the mixture gently will make fluffy matzoh balls; compressing the batter will make "cannonballs" — which many prefer. Cover the pot and boil gently for 30 to 40 minutes. The balls will fluff up and float to the top of the pot as they cook. The only way to tell if the matzoh balls are cooked through is to remove one from the water and cut it in half. It should be firm and uniform in color, with no wet, dark center. When the matzoh balls are done, remove from the pot with a spider or slotted spoon (they are too fragile to dump out into a colander).

4. Meanwhile, bring the broth to a boil in a soup pot. Add the greens and chicken, reduce the heat, and simmer, covered, until the greens are cooked through, 5 to 15 minutes, depending on the greens.

5. To serve, place two matzoh balls in each bowl and add a spoonful of the chicken and vegetables. Ladle the broth over all and serve immediately.

PASTA WITH GREEN CLAM SAUCE

Here's a quick "pantry meal" made with clam juice and canned clams. Adding fresh spinach to the clam sauce makes it a quick one-pot family supper that everyone loves. In the winter, frozen spinach can be substituted for the fresh.

Serves 4–6

- ¾ cup (5 ounces/155g) diced salt pork
- 4 garlic cloves, minced
- 1½–2 pounds fresh spinach, tough stems removed
- 1 (8-ounce) bottle clam juice
- 1 cup chicken broth (or substitute another 8-ounce bottle clam juice)
- 2 (6-ounce) cans chopped clams in clam juice
- ½ cup dried breadcrumbs
- 1 pound vermicelli
  Salt and freshly ground black pepper

1. Bring a large pot of salted water to a boil for the pasta.

2. Meanwhile, cook the salt pork in a medium saucepan over medium heat until the fat renders and the salt pork begins to crisp. Add the garlic and sauté until fragrant but not browned, about 30 seconds. Stir in the spinach; you may have to add the spinach a handful at a time, adding another handful as each addition wilts. Add the clam juice and broth. Strain the clam juice from the cans of clams into the saucepan. Add the breadcrumbs and let simmer for 10 minutes.

3. Cook the pasta in the boiling water until just al dente. Drain the pasta, reserving 1 cup of the pasta cooking water.

4. Return the pasta to the pot. Add the clams and the spinach mixture. Toss to mix. Add enough of the reserved pasta cooking water to make the pasta very moist. Season as needed with salt and very generously with pepper. Serve at once.

PASTA E FAGIOLI

This is such a warming bowl of comfort on a cold night. There is just enough meat in it to be satisfying. If you don't have salt pork, you can substitute bacon or pancetta, or cook a ham hock with the beans.

Serves 6

¾ cup (5 ounces/150g) diced salt pork

1 onion, diced

4 garlic cloves, minced

1 cup dried borlotti, pinto, red or white kidney beans, or great northern beans, soaked overnight and drained

1 teaspoon fine sea salt, plus more as needed

2 bay leaves

1 (28-ounce) can crushed tomatoes, or 1 quart tomato purée

1 tablespoon dried Italian seasoning, or a combination of dried oregano, rosemary, and thyme to equal 1 tablespoon

1–2 cups any poultry broth

1½ cups small pasta, such as rings or ditalini

Freshly ground black pepper

Freshly grated Parmesan cheese, for serving

1. Put the salt pork in a large soup pot and cook over medium heat until the fat renders and the little cubes are browned, about 8 minutes. Add the onion and garlic and sauté until limp, about 3 minutes. Add 8 cups of water, beans, salt, and bay leaves, and bring to a boil. Reduce the heat to a simmer, partially cover, and simmer until the beans are completely tender, 1½ to 2 hours. (The beans will not continue to soften once the tomatoes are added, so be sure they are fully cooked before continuing. I always sample at least 5 beans.)

2. Add the tomatoes and Italian seasoning and return to a boil. Simmer for 15 minutes.

3. Add enough of the broth to achieve a very soupy consistency (the pasta will thicken the soup) and return to a boil. Add the pasta and boil gently until the pasta is tender, about 20 minutes. Add salt and pepper to taste. Serve hot, passing the Parmesan at the table.

Pasture-raised pork is a little bit of a misnomer, since many pork farmers, particularly in the Northeast, are able to provide their pigs with acreage that includes forest (pigs love acorns and nuts!) as well as open areas. Their diet includes all sorts of tubers and grubs, and often vegetable leftovers, spent mash (Vermont has more beer brewers per capita than any other state), and whey (also lots of cheesemakers in the area). They get their iron naturally from the soil ("white meat" is anemic) and don't require vitamin and mineral supplements.

The meat from these animals is also darker because the animals get a lot of exercise as they forage for food. Muscle that's used more is darker because it is higher in myoglobin, which carries blood to the muscles. The age of an animal will also impact the myoglobin content of the muscles, with older animals having more myoglobin and darker meat. (By the way, the red liquid that leaks from most meat packages is myoglobin and not blood.)

The meat is likely to be richly marbled with fat and the roasts and chops will have lovely fat caps because pigs raised outdoors will need to regulate their body temperature with more fat. The pig's life outdoors, foraging and wallowing and playing, makes for a happy (or low-stress) environment, which translates into better-tasting meat.

The breed of the pigs makes a difference. Farmers turn to heritage breeds when they want to raise pigs that efficiently forage and convert their food to meat and fat. Where industrialized farms have developed breeds that produce "the other white meat" (lean, flavorless pork), heritage breeds tend to have well-marbled dark meat that is juicy with flavor.

Such meat doesn't need much enhancement. Season with salt and pepper, cook to 145°F, and let it rest for 10 minutes to redistribute the juices through the meat. The meat will be rosy in color and incredibly juicy.

RED BEAN GUMBO

Red beans and rice was such a favorite of Louis Armstrong, he signed his letters, "Red beans and ricely yours . . ." Put some of his music on to heighten your enjoyment of this Louisiana classic.

Serves 6–8

BEANS

2 cups dried red kidney beans, soaked overnight and drained

1 smoked ham hock

1 onion, quartered

1 teaspoon salt

3 bay leaves

GUMBO

½ cup (3.5 ounces/100g) any animal fat

½ cup unbleached all-purpose flour

2 onions, diced

2 celery stalks, diced

2 green bell peppers, diced

1 jalapeño, diced

4 garlic cloves, minced

8 ounces andouille or other spicy smoked sausage, thinly sliced

4 cups chicken broth

1 teaspoon ground white pepper

1 teaspoon dried thyme

Cayenne pepper

Salt and freshly ground black pepper

6 scallions, white and green parts, sliced

½ cup chopped fresh parsley

Freshly cooked rice, for serving

Louisiana-style hot sauce, for serving

1. To make the beans, combine the beans in a large pot with the ham hock, onion, salt, bay leaves, and at least 8 cups water, or enough to completely cover the beans and ham hock. Bring to a boil, then reduce the heat and simmer for 1½ to 2 hours. The beans should be completely tender; test at least 5 beans before deciding the beans are done. Remove from the heat and remove the ham hock from the pot. Set both aside.

2. To make the gumbo, melt the fat in a large heavy pot over medium-high heat. Whisk in the flour until it is completely incorporated into the fat. Reduce the heat to medium and stir constantly until the roux is at least medium brown in color, about 15 minutes. Do not let the roux burn or you will have to discard it and start over.

3. Add the onions, celery, bell peppers, and jalapeño, and stir until softened, 3 to 5 minutes. Stir in the garlic and sausage and sauté for 5 minutes. Stir in the broth. With a slotted spoon, transfer the beans from their cooking liquid into the gumbo. Season with the white pepper, thyme, and cayenne to taste. Bring to a rolling boil, reduce to a simmer, and cook for 45 minutes. While the gumbo cooks, remove the meat from the

ham hock, discarding any skin, the bone, and gristle, and add the meat to the pot. If the gumbo thickens too much, add some bean cooking liquid as necessary to maintain a soupy consistency.

4. Taste and adjust the seasoning with salt, black pepper, and cayenne. Add the scallions and parsley and serve immediately over rice, passing the hot sauce at the table.

SEAFOOD GUMBO

According to the *Picayune Creole Cook Book*, that famous historical record of Creole cooking originally published in 1901, it had become fashionable to use butter in place of lard in many recipes. But, the anonymous author noted in this classic, "It is a great fad among many of our day to use nothing but butter in cooking. The Creoles hold that butter should be used in its proper place, and lard in its own. The lard is not only less expensive, but is far preferable to an inferior quality of butter and, in many cases preferable to the best butter, according to the dish in course of preparation." Lard works great here, as do other animal fats.

Serves 6

6 tablespoons (2.6 ounces/75g) any animal fat
1 onion, diced
1 green or red bell pepper, diced
2 celery stalks, thinly sliced
1 pound okra, stems removed and pods sliced (about 4 cups)
4 garlic cloves, minced
6 cups chicken broth
8 ounces andouille or other spicy smoked sausage, thinly sliced
2 bay leaves
2 tablespoons fresh thyme leaves, or 1 tablespoon dried
½ teaspoon freshly ground black pepper, or as needed
½ teaspoon ground white pepper, or as needed
½ teaspoon cayenne pepper or hot or smoked paprika, or as needed
¼ cup unbleached all-purpose flour
1 pound shrimp, peeled and deveined
1 pound crabmeat, picked over for shells
Salt
2 scallions, white and green parts, chopped
¼ cup chopped fresh parsley
Freshly cooked white rice, for serving
Louisiana-style hot sauce, for serving

1. Melt 2 tablespoons of the fat in a large soup pot over medium-high heat. Add the onion, bell pepper, celery, okra, and garlic, and sauté until the onion is limp, about 4 minutes.

Recipe continues on next page

Seafood Gumbo, continued

2. Stir in the broth, sausage, bay leaves, thyme, black and white peppers, and cayenne. Bring to a boil, then simmer for about 30 minutes.

3. Meanwhile, make the roux. Melt the remaining 4 tablespoons fat in a skillet over medium-high heat. Whisk in the flour until you have a smooth paste. Reduce the heat and cook over medium heat, stirring, until the roux is a rich brown. This will take 15 to 30 minutes; do not let the mixture burn. If it does, you must throw it out and start over again. This is the darkened roux that gives gumbo its characteristic flavor and color.

4. Carefully stir the roux into the gumbo, taking care to avoid spatters. Add the shrimp and crab. Taste and adjust the seasonings, adding salt if needed. Simmer until the shrimp are pink and firm, about 15 minutes.

5. Remove the bay leaves. Stir in the scallions and parsley. To serve, ladle the gumbo over rice in large soup bowls. Pass the hot sauce at the table.

BATTUTO

Many of us are familiar with a mirepoix, the first step in making any number of French dishes, where aromatic vegetables, such as onions, garlic, celery, carrot, and perhaps a handful of fresh herbs, are finely chopped and sautéed in butter or olive oil. A similar mixture, a soffritto, sometimes with the addition of pancetta, starts many Italian dishes, while in Spain it's called sofrito. Perhaps less familiar is battuto. A battuto is a flavor base of finely chopped raw ingredients. The word is a derivative of *battere*, which means "to strike," and describes the use of a chef's knife chopping on a cutting board.

A battuto is usually a very finely chopped mixture of lardo (cured pork fatback), salt pork, lard, or pancetta with garlic and onions. It can also contain celery, carrots, peppers, chiles, and/or other chopped aromatic ingredients. In Italy, many butchers sell a premade basic battuto. Simmered in water, battuto can form the foundation of a meat broth. Lardo battuto is cured pork fat pounded to a cream with herbs and garlic.

NEW ENGLAND FISH CHOWDER

Whereas today much of the commercial salt pork is made from fatback or belly fat, salt pork in the good old days could be any cut (or trim) that was preserved in salt. After a typical slaughter in the fall, some pork was eaten fresh, some was smoked, some was turned into sausage, but most was salted and stored in barrels in the cool cellar or springhouse. This recipe is a classic from that era — made with either fresh fish (as here) or salted fish that has been freshened in several changes of water over several days.

Serves 4

4	ounces salt pork, diced
2	celery stalks, diced
1	onion, diced
1	carrot, diced
1½	pounds thin-skinned potatoes, diced
1	(8-ounce) bottle clam juice
3–4	cups chicken or fish broth
2	bay leaves
	Salt and freshly ground black pepper
1	pound white fish fillets, such as cod, pollock, haddock, or perch, any skin or pinbones removed
½–1	cup half-and-half, light cream, or heavy cream (optional; see Note)

1. Heat a large saucepan over medium heat and add the salt pork. Cook until the fat is fully rendered and the salt pork is crispy golden brown, about 8 minutes.

2. Add the celery, onion, and carrot to the pot and sauté, stirring occasionally, until the onion is softened but not browned, about 5 minutes.

3. Add the potatoes, clam juice, 3 cups of the broth, and the bay leaves. The broth should cover the potatoes. Add more broth if you want a soupier mixture. Increase the heat and bring to a boil. Cover, reduce the heat, and simmer until the potatoes are just cooked through, about 15 minutes.

4. With a potato masher, mash some of the potatoes to thicken the chowder. Reduce the heat to low and season with salt and pepper. Add the fish and cook over low heat for 5 minutes, then remove the pot from the heat and allow the chowder to sit, covered, for 10 minutes (so the fish will finish cooking but not overcook).

5. Remove the bay leaves and gently stir in the half-and-half, if desired. Taste and adjust the seasonings. Serve hot.

NOTE: Restaurants usually add heavy cream to their chowders. It makes the chowder deliciously rich and creamy. But the chowder is fine without any dairy at all. I usually add just ½ cup of half-and-half, to make it creamy but keep it light. You can skip the dairy entirely if you prefer.

FRIED RICE

In my days working in a Chinese restaurant, fried rice was the main dish of choice for a lot of male college students, and they didn't like sharing. This quantity can be served as a main dish or part of a multicourse Chinese meal. I've written it for maximum flexibility so that you can use whatever you have on hand. And yes, though salt pork has a strong tradition in colonial America, the Chinese also make salt pork and use it in soups and various stir-fries.

Serves 6–8

1	cup (7 ounces/200g) diced salt pork
1	onion, halved and thinly sliced vertically, or 1 leek, white and tender green parts, sliced
2	garlic cloves, minced
1	carrot, diced
4	cups shredded or thinly sliced green cabbage, napa cabbage, bok choy, or stemmed kale
4	cups packed cooked rice (see Note)
1–2	cups shredded or diced cooked meat or poultry
1	cup frozen peas
¼	cup Chinese oyster sauce
2	tablespoons soy sauce, plus more as needed
2	cups bean sprouts

1. Heat a wok over high heat. Add the salt pork and let brown, about 5 minutes. Add the onion and stir-fry until fragrant, about 1 minute. Add the garlic and carrot and stir-fry until the carrot begins to soften, about 3 minutes.

2. Add the greens and stir-fry until the greens are tender and wilted, about 5 minutes. Add the rice, cooked meat, frozen peas, oyster sauce, and soy sauce, and stir-fry until the rice is evenly colored and heated through and the vegetables are evenly distributed throughout the rice.

3. Add the bean sprouts and stir-fry for another minute, until well blended. Taste and add more soy sauce, if desired. Serve hot.

NOTE: To cook rice for this dish, combine 2 cups short-grain rice with 2½ cups water. Bring to boil in a covered saucepan, reduce the heat, and let simmer until all the water is absorbed, 10 to 15 minutes. Uncover and fluff with a large spoon.

SPAGHETTI ALLA CARBONARA

Eggs, bacon, cheese. These were familiar foods that GIs found to their liking in Rome during World War II. They brought the idea of the dish home with them and spread its popularity. Be generous with the black pepper; the *alla carbonara* part of the name translates as "in the style of coal makers," who presumably cooked over coals and had coal dust that fell into the dish. This dish is very rich and should be served with lots of vegetables on the side to balance it out.

Serves 3

- 4 ounces pancetta, salt pork, or bacon, cut into small cubes
- 2 garlic cloves, minced
- 4 egg yolks plus 1 whole egg
  Freshly ground black pepper
- ½ cup very finely grated Parmesan cheese
- ½ cup very finely grated pecorino cheese
- 9 ounces spaghetti
- 2 tablespoons chopped fresh parsley (optional)

1. Bring a large pot of salted water to a boil for the pasta.

2. Cook the pancetta in a large skillet over medium heat until it renders its fat and becomes crisp, about 8 minutes. Add the garlic and cook until fragrant, about 30 seconds. Reduce the heat to very low.

3. Meanwhile, whisk together the egg yolks and whole egg in a small bowl. Season well with pepper. Stir in half the cheeses.

4. Add the pasta to the boiling water and cook until al dente, 10 to 12 minutes. Drain the pasta, reserving 1 cup of the pasta cooking water.

5. Transfer the pasta to the skillet along with ½ cup of the pasta cooking water and stir to deglaze the pan, scraping up browned bits from the bottom with a wooden spoon. Stir to coat the pasta in the fat.

6. Pour the eggs over the spaghetti and continue to stir. The heat of the spaghetti and pan will cook the eggs, forming a thick yellow sauce. Sprinkle with the parsley, if desired, and serve with the remaining cheeses on the side.

PANFRIED BREADED FISH

On a fishing trip, the campers might save the bacon grease from the morning meal to fry up fish for dinner. But frying in lard or tallow? I've heard people express doubt, like you are somehow mixing metaphors, but it is not unusual to find fish-and-chip shops in England that fry in tallow. The truth is, lard and tallow allow foods to be fried without becoming greasy. This simple method of panfrying is easily adapted to other foods, including eggplant slices and chicken cutlets.

Serves 4

- 1 cup unbleached all-purpose flour
  Salt and freshly ground black pepper
- 2 eggs
- ¼ cup milk
- 1½ cups panko breadcrumbs or fresh white breadcrumbs
- 2 pounds skinless white fish fillets, such as cod, haddock, halibut, or perch
- ½ cup (3.5 ounces/100g) any animal fat, plus more as needed
  Lemon wedges, for serving
  Cocktail sauce or tartar sauce, for serving

1. Set up an assembly line on your counter with three shallow bowls. Put the flour in the first bowl and season generously with salt and pepper. Combine the eggs and milk in the second bowl and beat with a fork until well blended. Put the breadcrumbs in the third bowl. At the end of the assembly line, set up a sheet pan with a wire rack.

2. Dredge the fish fillets, one at a time, in the flour until well coated. Then dip in the egg mixture, then into the crumbs. Gently press the crumbs onto the fish on both sides. Transfer the fillets to the wire rack.

3. Heat the fat in a 12-inch cast-iron skillet over medium heat until hot. To test whether the fat is hot enough, sprinkle a few crumbs into the pan; when the fat is hot, the crumbs will sizzle immediately. Slide the fillets into the hot fat and cook until golden brown on both sides, 3 to 4 minutes per side. Do not crowd the pan. If you can't fit all the fillets in a single layer, work in batches. Return the cooked fish to the wire rack for 1 minute to drain. If you are working with thick fillets and are concerned that the crumbs have browned before the fish is cooked through, check the interior temperature of the fillets. Fish should be cooked to 145°F. If not sufficiently cooked through, bake in a hot (450°F) oven for a few minutes.

4. Serve immediately, with lemon wedges on the side. Pass the cocktail sauce at the table. (If you can't serve immediately, hold in a 200°F oven for up to 20 minutes.)

THE SCIENCE OF CRISP

When deep-frying or shallow-frying, the goal is always a crisp exterior and a moist interior. When food is slipped into the hot fat, moisture in the food starts to boil, turning into steam, pushing out to the surface of the food, preventing the hot fat from flowing in. At the same time, a starchy barrier on the food — be it a dusting of cornstarch or flour, breadcrumbs, or batter — also cooks, losing its moisture and drying into an impermeable shell.

Batters result in a smooth, crisp, and often delicate crust. Batter recipes vary widely, so results differ depending on ingredients. For example, high-gluten flours (like wheat) result in a chewy (some might say tough) crust, whereas gluten-free flours (like rice flour) result in a paper-thin ultra-crispy crust. Adding eggs or sugar to a batter will aid in browning, which may or may not be desirable. Batter coatings are smoother and have less surface area than most breaded coatings, so they tend to absorb less cooking fat. They also tend to offer the most protection for delicate foods, which is why fish is commonly battered before frying (think fish-and-chips).

Breading results in a crispy, crunchy, textured crust. Breaded coatings can range from fine breadcrumbs (think Chick-fil-A) to large, extra-crispy breading (think KFC, Popeye's, Japanese tempura). Fine breadcrumbs tend to absorb less fat than the extra-crispy style, since they provide less surface area for fat to soak into, but they are prone to sogginess. Extra-crispy-style breadings are usually achieved by incorporating large, already crispy particles, such as panko-style breadcrumbs, or cereal, such as cornflakes.

What makes the best coating? Batters don't allow as much steam to escape as breadcrumbs do, but they also prevent delicate foods like fish from falling apart. You can enhance the crispiness of batters by using rice flour or a finely ground cornmeal. Breadcrumb coatings stay crisp longer than batter coatings, but making the breadcrumbs stick to the food is a process. First dip the food to be fried in flour to dry its surface, then dip in an egg wash to help the crumbs stick. The more surface area the crumbs provide, the crispier the crust, so choose panko breadcrumbs rather than regular breadcrumbs; panko crumbs also don't compact the way regular breadcrumbs do, so more steam is allow to evaporate, which helps prevent the crust from getting soggy.

Once out of the fryer, the outside of the food is going to cool faster than the inside. The steam will keep flowing out, but this time it will collect just underneath the coating and turn the coating soggy. How to prevent this? First, hold the food in a spider or wire basket over the fryer for a few seconds to allow fat to drain off. Second, set the fried food on wire racks for good air circulation. And hold the food in a warm oven to help keep the outside crisp.

BEER-BATTERED FRIED FISH

This isn't a groundbreaking or original recipe, but it is included here as a reminder that everything is better when fried in an animal fat. For decades, tallow and lard were the frying media of choice in British fish-and-chip shops, until cheap oil and faulty health policy changed that practice. And no, the fat does not flavor the fish, though the fish does flavor the fat enough so you won't want to reuse it. Chose a thin, white-fleshed fish fillet; I'm not going to recommend a specific species because it is all about location. My fish of choice is Lake Champlain perch, caught by my brother.

Serves 4

    1 cup unbleached all-purpose flour
    1 cup cornstarch
    1 teaspoon salt
    1 teaspoon freshly ground black pepper
    1 (12-ounce) bottle or can beer (ale or lager)
    1 egg
 1½ pounds skinless fish fillets, ½ inch thick or
        less, cut into serving pieces if needed
    4 cups (1½ pounds/800g) lard or tallow
        Cocktail sauce, tartar sauce, and/or malt
        vinegar, for serving

1. Combine the flour, cornstarch, salt, and pepper in a large bowl and whisk until blended. Add the beer and egg and whisk until you have a smooth batter. Set aside for at least 30 minutes.

2. Pat the fish dry with paper towels. Add to the batter and turn to coat.

3. Heat the fat in a large deep skillet to 375°F. Set out wire racks on a large baking sheet to hold the cooked fish.

4. Turn the fish over in the batter to be sure each piece is well coated and slide into the fat. Add more than one piece if the pieces are small, but do not crowd the pan. Adjust the heat as needed to keep the temperature of the fat as close to 350°F as possible. Fry until the fish is golden brown, turning as needed, about 5 minutes per batch. Transfer the fish to the wire racks and allow the fat to return to 375°F before adding more fish to the fat. Continue until all the fish is fried.

5. Serve immediately, passing the cocktail sauce, tartar sauce, and/or malt vinegar at the table.

CHOOSE YOUR FRYING FAT WISELY. You will need a lot — 3 to 4 cups — which is why many people turned to cheap vegetable oil in the first place. Lard and tallow make a superior frying medium and result in crispier chicken. Poultry fats can also work, but it is hard to accumulate them in large quantities. Also, the smoke point of poultry fats is 375°F, which gives you less wiggle room than lard or tallow with their higher smoke points. Some people like to add bacon drippings to the lard or tallow for extra flavor.

SMALL PIECES WORK BEST. If using a whole bird, separate the legs from the thighs and cut the breasts into halves.

BRINE FIRST. Brining in buttermilk makes moister, more tender chicken. Don't skip this step. A 24-hour soak is best, but 4 hours works in a pinch.

SEASON GENEROUSLY. Colonel Sanders didn't get rich selling unseasoned chicken.

CHOOSE A HEAVY POT FOR EVEN COOKING. A deep skillet or Dutch oven works best. The pan will need a cover.

DON'T COOL DOWN YOUR FAT BY ADDING COLD CHICKEN. Let the chicken come to room temperature before breading and adding it to the fat.

USE A THERMOMETER TO TRACK AND MAINTAIN THE TEMPERATURE OF THE FAT. Melt the fat and heat it to a temperature of about 375°F. Add the chicken and the temperature should drop immediately. Aim to maintain a temperature between 325°F and 350°F; any lower and the chicken will be greasy, any higher and the crust will burn before the chicken is thoroughly cooked. It will be easier to maintain the proper temperature if you don't crowd the pan.

PLAN TO COOK IN BATCHES AND FINISH ON A RACK IN A HEATED OVEN. Each batch will take 15 to 18 minutes to fry, turning once. Finish off in a 400°F oven to guarantee the breast meat measures at least 150°F and the dark meat measures at least 165°F.

DRAIN AFTER FRYING. Drain on wire racks, not paper towels.

SPICY AND EXTRA-CRUNCHY SOUTHERN FRIED CHICKEN

This isn't Nashville-style hot, but it is nicely spiced. The chicken gets a tenderizing buttermilk bath beforehand, is twice rolled in a spiced flour coating, and is finished off in the oven — just to be sure it is cooked through. You can replace some of the lard or tallow with bacon fat for added flavor.

Serves 4–6

SPICE MIX

- 2 tablespoons sweet or hot paprika
- 1 tablespoon fine sea salt
- 1 tablespoon freshly ground black pepper
- 1 tablespoon ground white pepper
- 1 tablespoon garlic powder
- 1 tablespoon onion powder
- 2 teaspoons dried oregano
- ½ teaspoon cayenne pepper (optional)

CHICKEN

- 4 cups buttermilk
- 2 tablespoons sriracha hot sauce
- 1 (3½- to 4-pound) chicken, cut into 10 pieces (2 drumsticks, 2 thighs, 2 wings, 2 breasts, each breast cut in half)

COATING

- 1½ cups unbleached all-purpose flour
- ½ cup cornstarch
- 1 teaspoon baking powder

- 4 cups (1¾ pounds/800g) lard or tallow, for deep-frying

1. To make the spice mix, combine the paprika, salt, black and white peppers, garlic powder, onion powder, oregano, and cayenne (if desired) in a small bowl and mix thoroughly with a fork.

2. To marinate the chicken, combine the buttermilk, sriracha, and 2 tablespoons of the spice mix in a large bowl. Add the chicken pieces. The chicken should be completely submerged in the buttermilk. Refrigerate for 4 to 6 hours.

3. About 30 minutes before you plan to start cooking, set up a wire rack over a sheet pan to catch drips. Remove the chicken from the marinade and place on the wire rack to dry slightly. Reserve 1 cup of the marinade and discard the rest.

Recipe continues on next page

Spicy and Extra-Crunchy Southern Fried Chicken, continued

4. To make the coating, whisk together the flour, cornstarch, baking powder, and the remaining spice mixture in a large bowl.

5. Working with one piece of chicken at a time, put the chicken in the flour mixture and turn to coat, returning the coated chicken piece to the wire rack. Continue adding chicken pieces to the flour mixture one at a time until they are all coated and back on the rack.

6. Preheat the oven to 350°F and set up another sheet pan with a wire rack. Add the fat to a large skillet or Dutch oven and heat to 375°F over medium-high heat. Adjust the heat as necessary to maintain the temperature, being careful not to let the fat get any hotter.

7. Stir the reserved buttermilk mixture into the flour mixture to make it into a thick, wet batter. Starting with the dark meat pieces, add the floured chicken to the batter and turn to coat. Then add to the hot fat, skin side down. The temperature will drop, but try to bring it up to a steady 325° to 350°F temperature. Do not let the pieces touch, and work in two batches if needed. Fry the chicken until it's a deep golden brown on the first side, about 6 minutes. Carefully turn the chicken pieces with tongs or two spatulas and cook until the second side is golden brown, about 6 minutes longer.

8. Place the chicken on the lined sheet pan and bake for 5 to 10 minutes, until an instant-read thermometer inserted into the thickest part of the breast registers 150°F and the legs register 165°F. Serve hot or at room temperature.

MOCHIKO CHICKEN

Mochiko chicken, a dish you will find all over Hawaii, is probably adapted from *tatsuta age*, Japanese marinated fried chicken. When my son came home after WWOOFing (Willing Workers on Organic Farms) in Hawaii, he spoke so enthusiastically about the dish, I knew I had to make it for him. The only obstacle was finding the right flour — mochiko flour, or sweet rice flour, can be found online or where Asian foods are sold; no substitutes, please. Mochiko chicken is unique in that it is marinated in a batter flavored with soy sauce, garlic, and ginger and thickened with mochiko flour and cornstarch. My son tells me a dipping sauce is not needed, but I think a Thai sweet chili sauce is the perfect accompaniment.

Serves 4–6

3 pounds boneless, skinless chicken thighs
3 eggs, beaten
6 tablespoons soy sauce
¼ cup sugar
6 tablespoons mochiko flour (sweet rice flour)
6 tablespoons cornstarch, plus more as needed
4 scallions, white and green parts, finely chopped
4 garlic cloves, minced
   1-inch piece fresh gingerroot, peeled and minced
   Lard or tallow, for shallow-frying or deep-frying

1. Trim the fat from the chicken and cut into bite-size pieces. Reserve the fat for the next time you render chicken fat.

2. Beat the eggs in a large bowl. Add the soy sauce, sugar, flour, cornstarch, scallions, garlic, and ginger, and beat until well combined. Add the chicken and marinate for at least 4 hours, or up to 12 hours.

3. Take the chicken out of the refrigerator and mix well. The flour and cornstarch will have settled out of the batter. Mix vigorously and, if the batter no longer clings to the chicken, add a few more tablespoons of cornstarch to thicken it. Continue to mix the batter and chicken from time to time while you heat the fat for frying.

4. Heat 1 to 2 inches of fat to 375°F in a deep skillet. Set out a sheet pan with wire racks to drain the fried chicken.

5. Add the chicken, in batches, to the hot fat and cook until golden brown, 3 to 4 minutes per batch, turning once. Remove with a spider, drain for a few seconds over the fat, and then drain on the wire racks. Serve hot.

ROASTED CHICKEN THIGHS WITH ROOT VEGETABLES

Root vegetables roasted with poultry fat are sublime, and when you transform them into a one-dish supper, even better. Use whatever root vegetables you have on hand for this.

Serves 4

8	shallots, peeled and halved, or 1 onion, peeled and cut into wedges
6–8	garlic cloves, peeled and left whole
3	tablespoons (1.4 ounces/40g) any poultry fat
3–4	pounds bone-in chicken thighs
12	ounces thin-skinned potatoes, cut into 1-inch pieces
6	cups mixed peeled and cubed root vegetables (golden beets, carrots, celery root, parsnips, rutabagas, salsify, and/or turnips), cut into 1-inch cubes
2	teaspoons dried thyme
	Salt and freshly ground black pepper

1. Preheat the oven to 450°F.

2. Put the shallots and garlic in the center of a large roasting pan.

3. Heat a large skillet or Dutch oven over high heat. Add the fat and turn and tip the skillet until the fat is melted. Working in batches, add the chicken in a single layer, skin side down. Cook without stirring until the chicken skin is browned, about 5 minutes. Turn and cook on the second side for about 5 minutes. Remove the chicken from the pan and place on top of the shallots in the roasting pan. Repeat with the remaining chicken.

4. Add the potatoes and root vegetables to the skillet and toss with the fat. Transfer the vegetables to the roasting pan and arrange in a shallow layer around the chicken. Pour any excess liquid from the skillet over the vegetables and season the chicken and vegetables with the thyme, salt, and pepper.

5. Roast the chicken for 35 to 40 minutes, until the chicken registers 175°F on an instant-read thermometer.

6. Transfer the chicken to a serving platter, tent loosely with aluminum foil, and let rest. Meanwhile, toss the vegetables, spread them out, and return to the oven to continue to bake for about 10 minutes longer, until lightly browned. Transfer the vegetables to the platter with the chicken and serve.

CHICKEN AND BISCUITS

Chicken pie suppers are a fall-fundraising phenomena in Vermont. But what you get at these meals is usually chicken and biscuits, because when you bake the biscuits on top of the creamed chicken mixture, the biscuits suck too much moisture out of the sauce, which makes the dish too dry. And I don't limit myself to chicken in this dish; turkey and rabbit are terrific prepared this way. Use either leftovers from a roast or prepare the meat from scratch by poaching the meat (not boiling) in water with some aromatic vegetables (onion, carrot, celery, or celery root) for 20 to 45 minutes. Change the vegetables with the seasons. Poultry fat is the best choice for the biscuits, and any poultry fat will enrich the sauce.

Serves 6

Buttermilk Biscuits (page 214; prepared up to step 3)

4 cups fresh or frozen cubed or chopped vegetables, peeled if necessary

Fine sea salt

6 tablespoons (2.6 ounces/75g) any poultry fat

2 shallots, minced, or 1 leek, white and tender green parts, thinly sliced

2 garlic cloves, minced (optional)

6 tablespoons unbleached all-purpose flour

4 cups chicken broth

4 cups cooked chicken (or turkey or rabbit)

1 teaspoon dried thyme, or 2 teaspoons fresh

Freshly ground black pepper

1. Prepare the biscuits according to the directions on page 215, up through step 3. Place in the refrigerator.

2. Preheat the oven to 450°F.

3. If you are using fresh root vegetables, place in a saucepan, cover with water, and add about 2 teaspoons fine sea salt. Bring to a boil and boil until just tender, about 10 minutes. Drain. If you are using fresh summer vegetables, steam over boiling water until tender, about 5 minutes. If you are using frozen vegetables, remove from the freezer.

4. Melt the fat in a large saucepan over medium heat. Add the shallots and garlic, if desired, and sauté until fragrant and limp, 3 to 5 minutes. Sprinkle in the flour and stir until all the flour is wet. Whisk in the broth and stir until thickened and smooth. Stir in the chicken, vegetables, and thyme. Taste and adjust the seasoning with salt and pepper. Bring to a simmer. Keep hot while you bake the biscuits.

5. Remove the biscuits from the refrigerator. Bake according to the recipe directions for 15 to 18 minutes, until the biscuits are golden.

6. To serve, split open one or two biscuits for each serving. Ladle the chicken and vegetable mixture over the biscuit halves and serve immediately.

CHICKEN AND DUMPLINGS

Chicken and dumplings is an essential comfort food of the Pennsylvania Dutch and folks in the American South and Midwest. It is a beautifully efficient dish, taking one chicken to make the meat, the broth, and, in this version, the fat in the dumplings (other versions use lard or butter). Part of the popularity of this dish stems from the way a single chicken can be stretched to feed a crowd by making more dumplings. I add more vegetables than usual because I like to turn this dish into a one-pot dinner. If you prefer, you can use chicken parts instead of a whole chicken. Please note that the broth and chicken should be made a day ahead and chilled to allow the chicken fat to solidify.

Serves 6–8

BROTH

- 1 (3- to 5-pound) chicken, or 3–5 pounds chicken parts, preferably dark meat
- 1 onion, quartered
- 2 celery stalks, chopped
- 1 bunch parsley

DUMPLINGS

- ¼ cup (1.4 ounces/50g) chicken fat (skimmed from the broth)
- 3 cups unbleached all-purpose flour
- 1 teaspoon baking powder
- 1 teaspoon fine sea salt, plus more as needed
- 1 teaspoon dried thyme
- 1 cup milk or water
  Freshly ground black pepper

VEGETABLES

- 3 carrots, diced
- 3 celery stalks, diced
- 1 cup frozen peas
- 1 cup frozen cut green beans

1. To make the broth, put the chicken in a large soup pot. Add enough water to cover completely, then add the onion, celery, and half of the parsley, reserving the remaining parsley for a garnish. Cover the pot, set over high heat, and bring to a boil. Reduce the heat to low, partially cover, and very gently simmer until the chicken is just cooked through, 1 to 2 hours (depending on whether you are cooking parts or a whole bird).

2. Strain the broth and chill overnight. Discard the vegetables in the broth and chill the chicken.

Recipe continues on next page

3. The next day, to make the dumplings, line a sheet pan with parchment paper or a silicone mat. Skim the fat from the chicken broth and measure. Add additional rendered poultry fat to make the ¼ cup needed. If there is more than ¼ cup, refrigerate the remainder to use in another recipe within 1 week, or freeze.

4. Combine the flour, baking powder, salt, and thyme in a large bowl and whisk to blend. Add the chicken fat and stir to coat the flour, then add the milk. Stir just enough to combine, then knead until all the flour is incorporated into the dough.

5. Turn the dough out onto a lightly floured work surface and divide in half. Take the first portion and roll out into a thin rectangle. Using a pizza cutter or sharp knife, slice the dough lengthwise into 1-inch-wide strips, then cut crosswise into pieces about 2 inches long. Repeat with the remaining dough. Transfer the dumplings with a metal spatula or pancake turner to the prepared sheet pan and allow to rest for 30 minutes.

6. Transfer 7 cups of the broth to a large pot and reserve the rest for another use (use within 1 week, or freeze). Begin heating the broth over very low heat. Separate the chicken meat from the bones and discard the bones and skins. Finely chop the remaining parsley, discarding the stems.

7. To cook the dumplings, bring the broth to a boil. Add salt and pepper to taste. Drop the dumpling strips into the boiling liquid, along with the diced carrots and celery. Reduce the heat to a simmer and cook, uncovered, stirring occasionally to prevent sticking, until the dumplings are tender and slightly puffed, the broth is slightly thickened, and the carrots are tender, about 30 minutes.

8. Return the chicken to the pot, add the peas and green beans, and cook until heated through, 5 to 10 minutes. Stir in the reserved parsley and adjust the seasoning as needed. Serve hot.

CHICKEN PAPRIKASH

Lard is the fat most typically used in Hungary, even when the dish in question is chicken. In this classic dish, the chicken is browned in lard, then braised in an orange-red sauce of sweet paprika and chicken broth. The paprika must be fresh! If your paprika is more than a few months old, buy fresh for this dish. And before you toss out the old stuff, compare the colors. You'll see what a difference a few months make — with the fresh paprika a bright orange and the older paprika a dull brick red.

Serves 4–6

- ¼ cup (1.7 ounces/50g) lard (or any animal fat)
- 1 (3- to 4-pound) chicken, cut into 8 pieces, or 4 pounds bone-in chicken thighs
- Salt and freshly ground black pepper
- 1 large onion, diced
- 3 Hungarian hot wax peppers, seeded and diced
- 1 green bell pepper, seeded and diced
- 3 garlic cloves, minced
- 3 tablespoons unbleached all-purpose flour
- 3 tablespoons Hungarian sweet paprika
- 2 cups chicken broth
- 1 (14.5-ounce can) diced tomatoes, or 1½ cups seeded and diced fresh tomatoes
- ½ cup sour cream
- Freshly cooked egg noodles, for serving

1. Melt the lard in a large Dutch oven over medium-high heat. Generously season the chicken with salt and pepper. Working in batches, partially cook the chicken, turning once, until browned, 8 to 10 minutes per batch. Transfer the chicken to a bowl; set aside.

2. Add the onion, hot wax peppers, and bell pepper to the Dutch oven and cook, stirring occasionally, until soft, about 8 minutes. Add the garlic, flour, and paprika, and cook, stirring, for about 2 minutes, to cook out the raw flavor of the flour.

3. Add the broth and tomatoes to the Dutch oven and then return the chicken and its juices to the pot. Bring to a boil, reduce the heat to medium-low, and simmer, partially covered, until the chicken is fully cooked, about 30 minutes.

4. Remove the pot from the heat and let stand for 5 minutes. Add a spoonful of the sauce to the sour cream in a small bowl and mix well, then stir the sour cream into the Dutch oven. This step tempers the sour cream and makes it less likely to break or curdle.

5. Make a bed of egg noodles on a serving platter or individual plates. Set the chicken on the egg noodles, then ladle the sauce over the chicken and serve.

DEEP–DISH CHICKEN PIE

There are many steps to this showstopper pie, but the steps aren't all that time-consuming. If you make this a day ahead, then reheat, the filling will hold together and cut into neat slices. But any way you serve this, it is delicious.

Serves 6

PASTRY

- 3¾ cups unbleached all-purpose flour
- 2 teaspoons fine sea salt
- 1¼ cups (8.7 ounces/250g) any fat (except bacon grease; chilled poultry fat is recommended)
- ¾ cup very cold water
- 2 eggs

CHICKEN

- 1½ pounds boneless, skinless chicken thighs
- 2 celery stalks, halved
- 1 onion, quartered
- 4 cups water
- 1 teaspoon fine sea salt

FILLING

- 3 cups diced mixed root vegetables (golden beets, carrots, celery root, parsnips, rutabagas, salsify, and/or turnips)
  Salt
- 6 tablespoons (2.6 ounces/75g) any animal fat (poultry fat is recommended)
- 4 garlic cloves, minced

- 2 shallots, diced
- ½ cup unbleached all-purpose flour
- 1 teaspoon poultry seasoning
- 1 teaspoon dried sage
  Freshly ground black pepper

1. To make the pastry, whisk together the flour and salt in a bowl. If you are working with poultry fat or lard, add the fat and rub it into the flour with your hands until the fat is in pieces about the size of peas. If you are using tallow, combine the tallow, salt, and half of the flour in a food processor. Process until the mixture is evenly blended into little pebbles. Add the remaining flour and process until well blended. Transfer the mixture to a bowl.

2. Measure the water into a glass measuring cup, add the eggs, and beat to combine. Add about half of the mixture to the flour mixture and stir with a fork to combine. Add more of the egg mixture as needed to get the dough to come together into a ball. It should not be wet or sticky. Reserve the remaining egg mixture for an egg wash.

3. Divide the dough into two unequal pieces, with two-thirds of the dough for the bottom crust and one-third for the top crust. Form each piece into a disk, wrap in plastic wrap, and chill for at least 30 minutes in the refrigerator, up to 1 day.

4. To make the chicken, combine the chicken, celery, and onion in a saucepan; cover with the water and add the salt. Bring to a simmer over medium heat and simmer until the chicken is mostly cooked, about 10 minutes. Remove from the heat and let the chicken cool in the broth.

5. Lightly grease a 9-inch springform pan and place it on a sheet pan lined with parchment paper to aid in cleanup.

6. On a floured surface, roll out the larger piece of dough to a thickness of ¼ inch. Carefully transfer it to the springform pan, tucking it into the bottom edge, and trim the top edge so there is a ½-inch overhang all around. Prick the dough with a fork on the base and sides. Place in the freezer to chill the dough for at least 30 minutes.

7. Preheat the oven to 425°F with a rack in the lower third of the oven.

8. While the bottom piecrust is chilling, begin making the filling. Put the root vegetables in a small saucepan, cover with water, add salt generously, and bring to a boil over medium-high heat. Reduce the heat to maintain a simmer and cook until the vegetables are tender, about 10 minutes. Drain and set aside.

9. Partially bake the pastry for about 20 minutes, until the crust is beginning to brown at the edges. Set aside to cool to room temperature and reduce the oven temperature to 350°F.

10. Meanwhile, drain the cooled chicken, reserving the broth and discarding the vegetables. Cut the chicken into bite-size cubes.

11. Melt the fat in a saucepan over medium heat. Add the garlic and shallots and sauté until fragrant, about 1 minute. Whisk in the flour. Cook until the mixture is just starting to color, 2 to 3 minutes. Gradually whisk in 2½ cups of the broth (reserve the remaining broth for another use). Bring the mixture to a simmer and let cook until thickened, about 5 minutes. Add the drained root vegetables, chicken, poultry seasoning, and sage. Add salt and pepper to taste. Remove from the heat.

12. Spoon the filling into the cooled bottom crust, pressing down firmly to release any air pockets.

Recipe continues on next page

13. Roll out the remaining piece of dough on a lightly floured surface to a thickness of ¼ inch. Use a rolling pin to transfer the dough to the top of the pie. Trim away any excess dough, leaving ¾ inch all the way around the edge. Nudge the edge down into the side of the springform pan so that it meets the top edge of the bottom crust. Push the crust down a bit so the excess puckers outward and creates a little lip. Press that outer lip together to seal the edges a bit, then crimp with a fork to seal. Brush the reserved egg wash evenly over the top crust and cut steam vents into the top crust. Transfer the pie to the prepared sheet pan.

14. Bake for about 1 hour 15 minutes, until the crust is very golden and the filling is bubbling through the vents.

15. Cool the pie for at least 45 minutes before slicing and serving; the longer it sits, the neater the slices will be. Remove the outer edge of the springform pan. Slide an offset spatula around the edge of the base of the pan; if the pie is really cool and it feels solid in the middle, you'll likely be able to pick up the pie with your hands (or a large spatula) and transfer it to a platter or stand. If it feels soft in the middle, you may rather keep it on the springform base for slicing.

HAINANESE CHICKEN RICE

This is one of those recipes where the end result is much, much greater than the sum of its parts. Hainan is China's smallest and most southern province, with a tropical climate similar to Malaysia's. Hainanese chicken rice is a dish you'll find in any Southeast Asian country where there are Chinese immigrants, including Malaysia, Vietnam, Thailand, and Singapore. The basic method is universal: Poach a chicken, then use the liquid to make rice. In Vietnam, the cooking liquid might have fish sauce and lime leaves; in Malaysia, it might have pandan leaves and lemongrass.

Yes, it is a bit fussy to make. And no, this isn't the most streamlined version you can find, but most of the streamlined recipes I've looked at focused on not using the "notoriously" fatty skin and replacing the chicken fat with vegetable oil, both of which would diminish the flavor.

You will end up with more broth than you need for the rice and accompanying sauces. The extra broth can be served on the side, garnished with chopped scallion. Or you can reserve the broth for another use, or you can do what some restaurants do, which is save the broth to replace the water in the recipe the next time they make it, resulting in a broth that gets more and more intense each time you use it.

Serves 6–8

CHICKEN

8 garlic cloves, peeled

4-inch piece fresh gingerroot, peeled

1 (6- to 8-pound) chicken

2 lemongrass stalks, smashed with the side of a knife (optional)

CHILI SAUCE

2 tablespoons sugar

¾ cup chicken broth (reserved from above)

¼ cup freshly squeezed lime juice

1 tablespoon fish sauce

1 cup fresh cilantro leaves, finely chopped

3 scallions, white and green parts, finely chopped

1–3 fresh red or green chiles, finely chopped (seeding is optional)

6 garlic cloves, finely chopped

1-inch piece fresh gingerroot, peeled and finely chopped

Salt

SOY DRIZZLING SAUCE

½ cup sweet soy sauce

½ cup soy sauce

¼ cup oyster sauce

1 cup chicken broth (reserved from above)

2 scallions, white and green parts, finely chopped

4 garlic cloves, minced

RICE

Chicken fat and skin (reserved from above)

2 cups long-grain white rice

Salt

4 scallions, white and green parts, finely chopped

1 cup fresh cilantro leaves, finely chopped

1. To make the chicken, smash four cloves of the garlic and finely chop the remaining four cloves. Slice the ginger. Now finely chop half of the ginger slices. Set aside the chopped garlic and ginger.

2. Reach into the cavity of the bird and remove any chicken fat and set aside; do not discard. Stuff the cavity of the chicken with the four smashed garlic cloves, the ginger slices, and the lemongrass (if desired). Place the chicken in a pot just large enough to hold it and cover with water.

3. Bring the chicken to a simmer over medium heat and simmer until the chicken registers 185°F in the thickest part of the thigh, about 2 hours. Remove from the heat and let the chicken cool in the broth.

Recipe continues on next page

Hainanese Chicken Rice, continued

4. While the chicken cools, make the sauces. To make the chili sauce, combine the sugar and ¼ cup of the broth in a heatproof bowl and stir to dissolve. (If the broth isn't hot enough to melt the sugar, give it a quick zap in the microwave.) Add the remaining ½ cup broth, lime juice, fish sauce, cilantro, scallions, chiles, garlic, and ginger. Add salt to taste. To make the soy drizzling sauce, combine the sweet soy sauce, regular soy sauce, and oyster sauce in a small bowl. Add ½ cup of the broth, the scallions, and garlic. Taste and add as much of the remaining ½ cup broth as you like. Set both sauces aside.

5. When the chicken is cool enough to handle, remove all the skin; it should pull off easily. Set the skin and chicken aside. Reserve the broth.

6. To make the rice, finely chop the reserved chicken fat and the skin you just removed. Place a large heavy saucepan over medium heat and add the fat and skin. Cook gently until liquid fat covers the bottom of the pot, 5 to 10 minutes, stirring almost constantly to prevent sticking. Add the rice and reserved chopped garlic and ginger. Stirring constantly, toast the rice until well coated with the chicken fat and appearing almost dry. Add 4 cups of the reserved broth and a generous pinch of salt, reduce the heat to low, cover, and cook until the rice has absorbed all the liquid and is cooked throughout, 18 to 20 minutes. Refrigerate any remaining broth for another use.

7. While the rice cooks, carve the chicken into serving-size pieces, on the bone or off.

8. When the rice is done, fluff with a fork. Stir in the scallions and cilantro. Taste and add salt if needed, remembering that the soy drizzling sauce is salty.

9. To serve, mound the rice on individual plates or a large platter and top with the chicken. Drizzle some of the soy drizzling sauce over the chicken and serve, passing the remaining drizzling sauce and the chili sauce at the table. Provide each diner with small bowls or ramekins for the sauces.

FRENCH WHITE BEAN AND CABBAGE SOUP WITH DUCK CONFIT

When I prepare duck, I try to get as many meals from the bird as possible because ducks are not cheap. This rib-sticking French *garbure* (soup or stew) makes use of the duck broth made from the breast and backbones. The small amount of duck confit needed for the garnish comes from the duck wings. All that fine duckiness adds flavor, but even if you don't have duck confit on hand and must use chicken broth, it is a soup worth making. The salt pork and ham add plenty of flavor.

Serves 6–8

- 4 ounces salt pork, diced
- 1 pound any garlicky sausage, such as Italian sausage, chopped
- 2 leeks, white and tender green parts, sliced
- 2 carrots, diced
- 4 garlic cloves, minced
- 2 cups dried white beans (cannellini, navy, or great northern), soaked overnight and drained
- 10 cups duck or other poultry broth, or a mixture
- 1 smoked ham hock
- 2 bay leaves
- 1 rosemary sprig
- 1 pound thin-skinned potatoes, peeled and diced
- 4 cups thinly sliced savoy or green cabbage
  Salt and freshly ground black pepper
  Duck Confit (page 261), for garnish

1. Cook the salt pork in a large Dutch oven over medium heat until the fat renders out and the salt pork becomes crisp, about 8 minutes. Add the sausage, leeks, carrots, and garlic, and sauté until the vegetables are softened and fragrant and the sausage is mostly cooked, about 5 minutes. Add the beans, broth, ham hock, bay leaves, and rosemary, and bring to a boil. Reduce the heat and simmer, uncovered, until the beans are tender, 1 to 1½ hours.

2. Add the potatoes and cabbage and simmer for 1 hour.

3. Taste the soup and season with salt, if needed, and pepper.

4. Chop or shred the duck confit; how much to use is up to you. Serve the soup hot, garnished with a little duck confit.

DUCK CONFIT WITH POTATOES

When I was younger, I traveled in France on something like $5 a day. I have to admit I wondered why people made such a fuss over duck confit. It wasn't until I made my own that I learned how good it can be — in fact, it is one of the very best ways to enjoy duck. And that confit I had in my youth? I suspect it was old — very, very old. This is traditionally served with a bitter green dressed with a sharp vinaigrette; braised kale or cabbage is also a great accompaniment.

Serves 4

2 pounds thin-skinned potatoes, peeled and cut into ½-inch cubes or wedges

Salt

4 Duck Confit legs (page 261)

¼ cup (2.4 ounces/50g) duck fat

¼ cup minced fresh parsley

Freshly ground black pepper

1. Preheat the oven to 450°F with a rack in the top third of the oven. Line a sheet pan with aluminum foil.

2. Put the potatoes in a large pot, cover with cold water, and add salt generously. Bring to a boil over high heat. As soon as a full boil is reached, test for doneness — the potatoes should be just tender when pierced with a knife tip. If the potatoes aren't tender, reduce the heat to a simmer and cook over medium heat until done; it should take just a few minutes. Immediately drain the potatoes and spread them on paper towels to dry thoroughly.

3. Remove the duck legs from the fat and use a silicone spatula to gently scrape off any excess fat that clings to the skin. Arrange the duck legs, skin side up, on the prepared sheet pan. Roast for 20 to 25 minutes, until the skin is a deep golden brown and very crisp, checking often near the end to prevent burning.

4. While the duck is in the oven, melt the duck fat in a large heavy skillet over medium-high heat. When hot, add the potatoes and cook until deep golden and crisp, about 20 minutes, turning occasionally. When ready to serve, add the parsley and toss gently. Taste and season with salt and pepper — it may not need any. Serve hot with the crisp duck legs.

SMOTHERED PORK CHOPS

Who doesn't love gravy? "Smothered" dishes come to us from the South, as a way of cooking meats that don't create their own gravies. It involves browning the meat, then simmering it in an onion gravy. Traditionally, this is served over rice, but it is also pretty tasty with mashed potatoes. The brining step is optional, especially with pasture-raised pork, but it does enhance the dish.

Serves 4

BRINE AND CHOPS

  3 cups water

  ¼ cup fine sea salt

  ¼ cup firmly packed light brown sugar

  2 cups ice cubes

  3 bay leaves

  1 teaspoon allspice berries

  1 teaspoon black peppercorns

  4 bone-in pork chops, ¾–1 inch thick

SEASONED FLOUR

  ½ cup unbleached all-purpose flour

  2 teaspoons fine sea salt

  2 teaspoons garlic powder

  1 teaspoon onion powder

  1 teaspoon smoked paprika

  ½ teaspoon freshly ground black pepper

  ¼ teaspoon ground white pepper

GRAVY

  ½ cup (3.5 ounces/100g) any animal fat
     (lard is recommended)

  2 onions, thinly sliced

  2 cups pork or chicken broth

  ½ cup buttermilk

1. To make the brine, stir together 1 cup of the water and the salt and brown sugar in a glass measuring cup or a saucepan. Microwave for 1 minute or heat over high heat and stir just until the salt and sugar dissolve. Stir in the additional 2 cups water, the ice cubes, bay leaves, allspice, and peppercorns, and cool to 45°F or lower. (You don't want to cook the chops in the brine.)

2. Put the pork chops in a glass baking dish or resealable plastic bag. Pour in the chilled brine and cover the dish or seal the bag. Refrigerate for about 2 hours. Remove the chops, discard the brine, and pat the chops dry.

3. To make the seasoned flour, mix together the flour, salt, garlic powder, onion powder, paprika, and the black and white pepper in a shallow bowl. Dredge each chop in the flour to coat. Set the chops on a wire rack and reserve the excess seasoned flour.

Recipe continues on next page

4. Melt the fat in a large heavy skillet over medium-high heat. When the fat is hot, add the chops and brown on each side for about 3 minutes. Remove the chops from the skillet.

5. To make the gravy, reduce the heat to medium, add the onions, and cook until golden, about 5 minutes. Add the reserved seasoned flour to the skillet and cook until the flour is toasted to a golden brown, stirring constantly, about 5 minutes. Stir in the broth and bring to a boil. Reduce the heat to a simmer and stir in the buttermilk. Taste and adjust the seasoning. Return the chops to the skillet and simmer until the chops are cooked to 140°F, 5 to 15 minutes, depending on the thickness of the chops. Do not overcook or the chops will be dry. Serve hot.

BISCUITS AND SAUSAGE GRAVY WITH LEEKS

Biscuits and sausage gravy, at best, is a hearty breakfast. At its worst, it is a gloppy mess, indifferently slopped onto hard, stale biscuits. Can I be forgiven for making this Southern classic a wee bit lighter and more flavorful with the addition of a leek and a light hand with flour? You can lighten it further with more vegetables, such as frozen peas or green beans. I make the sausage from scratch, but you can substitute your favorite breakfast sausage, if you prefer.

Serves 6

BREAKFAST SAUSAGE

- 1 pound ground pork or other ground meat, such as ground turkey
- 2 tablespoons minced onion
- 2 tablespoons pure maple syrup
- 2½ teaspoons dried sage
- 1 teaspoon fine sea salt
- ½–¼ teaspoon dry mustard
- ¼ teaspoon freshly ground black pepper
- ⅛ teaspoon ground cloves

GRAVY

- 1 tablespoon (0.45 ounce/13g) lard
- 1 leek, white and tender green parts, halved and sliced
- ¼ cup unbleached all-purpose flour

2½ cups whole milk

    Salt and freshly ground black pepper

    Buttermilk Biscuits (page 214)

1. To make the sausage, combine the pork in a bowl with the onion, maple syrup, sage, salt, dry mustard, pepper, and cloves. Mix well with your hands until the seasonings are evenly distributed. Form a small patty of the mixture and cook in a large skillet over medium-high heat until browned. Taste and then adjust the seasoning in the rest of the mixture as needed.

2. To make the gravy, melt the lard in the large skillet over medium-high heat, add the sausage mixture, and cook until evenly browned, about 8 minutes. Add the leek and sauté until translucent, about 3 minutes. Sprinkle the flour over the sausage. Stir it in thoroughly, coating the meat and leek. Let cook for about 3 minutes, stirring constantly to prevent burning. Stir in the milk and bring to a low simmer, stirring. Continue to simmer for about 3 minutes, stirring occasionally and adjusting the heat as needed. Taste and season with salt and pepper.

3. To serve, split open the biscuits and ladle the gravy over.

WHERE'S THE FLAVOR?

This book is admittedly meat-centric. It should go without saying that every mention of poultry, pork, beef, and fat is referring to pasture-raised animals, preferably from heirloom breeds that were raised on pasture, not in modern feedlots.

Pigs used to be raised to have extra fat on them, for use as rendered lard, salt pork, and bacon. According to *Modern Farmer*, in the 1940s, North American pigs had 2¾ inches of fatback on them; as of 2006, they had an average of less than ¾ inch. Responding to customer demand, from 1980 to 2000, pork breeders in America succeeded in reducing the fat in pigs by over 30 percent. From the 1950s to 2000, it was reduced by 75 percent overall. The loss of fat has meant a loss of flavor.

All animals put down fat in the same general pattern. First, fat is allowed to accumulate around the internal organs to protect them from the bumps and bangs of life. Then fat accumulates under the skin, then it is deposited intramuscularly, as marbling. A pasture-raised animal is slower growing than a grain-fed animal, so it accumulates fat in a slower pattern.

When buying meat, don't just look for "grass-fed" or "pasture-raised." Ask for animals that are older; these will have the better, more intense flavor.

PORK CARNITAS

Carnitas, which translates from Spanish as "little meats," is the Mexican version of pulled pork — or maybe duck confit. The pork is simmered in lard until falling-apart tender, then fried until crispy. Pork carnitas can be served as a filling for tacos, burritos, or tamales. Leftovers can be made into hash, turned into a barbecue sandwich with the addition of barbecue sauce, or just served alongside rice, beans, and pico de gallo.

Serves 8

- 1 (3-pound) boneless or (6-pound) bone-in pork shoulder, cut into 2-inch pieces
- 1 onion, quartered
- 1 orange, quartered
- 1 lime, quartered
- 1 jalapeño, chopped
- 8 garlic cloves, chopped
- 1½ teaspoons fine sea salt
- 1½ teaspoons ground cumin
- 1 teaspoon dried oregano
- ½ teaspoon freshly ground black pepper
- ¼ teaspoon ground chipotle chile
- ½ cup (3.5 ounces/100g) lard

1. Preheat the oven to 250°F.

2. Put the pork in a large Dutch oven, with or without the bone (the bone can be saved for soup if you prefer). Tuck the onion, orange, lime, jalapeño, and garlic among the pieces of pork. Mix together the salt, cumin, oregano, black pepper, and ground chipotle in a small bowl. Sprinkle over the meat. Spoon the lard over all, dropping dollops of lard evenly on top of the pork. Cover with the lid.

3. Bake for about 6 hours, until the meat is meltingly tender and bathed in its own juices.

4. Use a slotted spoon to remove the meat and set aside in a bowl. Strain the juices into a fat separator or glass measuring cup.

5. Remove the meat from the bone, if necessary, and shred the meat with two forks. Separate the fat from the juices. You should have about ½ cup fat and about 1 cup of drippings.

6. To serve, pour enough fat into a large skillet to cover the bottom, about ¼ cup, and heat over high heat. Working in two or three batches, add enough meat to cover the bottom in a shallow layer and fry until crispy on the first side, about 5 minutes, and 3 minutes on the second side; do not stir the meat as it fries. Transfer to a bowl and keep warm. When the last batch is crisp, return all the meat to the skillet, add the reserved juices, and stir to evenly moisten the meat. Remove from the heat and serve.

It has always amazed me when I find recipes for gravy that start with a roux made with butter or oil. Doesn't everyone understand that the flavors of the meat or bird are found in the fat? That fat should never be discarded.

After you cook a roast, be it meat or fowl, transfer the roast to a platter to rest under a tent of aluminum foil. Pour the liquid — a combination of juices and fats — into a fat separator or glass measuring cup. Place the roasting pan over a burner or two and deglaze the pan with 1 cup of broth, wine, or water over medium-high heat, scraping up any burnt bits from the bottom of the pan. Pour the resulting liquid into the fat separator. Within 5 minutes, most of the fat will have risen to the top, where it can be spooned off.

When you are ready to make the gravy, heat 2 tablespoons of the separated fat and whisk in 2 tablespoons of flour for every 1 cup of gravy (using the pan drippings plus any broth extender you might need) in a saucepan over medium heat. If you are making a lot of gravy and don't have enough fat, extend the fat with rendered fat (using any poultry fat for any roasted bird, lard for a pork roast, and tallow for a beef roast) or butter. Stir in an equal amount of unbleached flour, whisking for 1 or 2 minutes until the flour smells toasted. Stir in the pan drippings and broth and bring to a boil. Reduce the heat to a simmer, season to taste, and serve hot.

Reserve any leftover fat from the roast to cook with or to serve as drippings on toast and use within the week, or freeze for long-term storage. This fat will be softer than the fats you render from solid chunks of leaf tallow or lard.

CURRIED BEEF PASTIES

Whether you call them patties (as in Jamaican beef patties) or pasties, beef is a popular filling, and curry is a popular flavoring. This hot pocket nods to Jamaica, with the addition of curry powder in the dough, but the style of filling is more like a British pasty, with diced beef and a mild flavor. Jamaican beef patties should be made with Scotch bonnet chiles, which are too hot for my palate.

Makes 12 pasties

   2 tablespoons (0.8 ounce/25g) tallow or
       another animal fat
  12 ounces stew beef (not lean), finely chopped
   1 onion, finely chopped
   2 tablespoons curry powder
   2 teaspoons fine sea salt, plus more as needed
   8 ounces thin-skinned potatoes (peeling is
       optional), finely chopped
   1 carrot, finely chopped
   1 cup water
   2 tablespoons ketchup or tomato paste
   2 tablespoons unbleached all-purpose flour
       Freshly ground black pepper
       Cayenne pepper (optional)
  ¼ cup frozen peas, thawed
       Pasty Pastry (page 273), made with 1 table-
       spoon curry powder mixed into the flour
   1 egg beaten with 1 tablespoon water, for
       egg wash

1. Melt the tallow in a large skillet over medium-high heat. Add the beef, onion, curry powder, and salt, and cook, stirring occasionally, until the meat is browned, about 5 minutes. Add the potatoes and carrot and cook, stirring occasionally, until the potatoes are translucent, 3 to 5 minutes longer. Add ¾ cup of the water and the ketchup and cook, stirring and scraping up any browned bits from the bottom of the skillet, until the potatoes are tender, about 5 minutes.

2. Combine the flour with the remaining ¼ cup water and stir until no lumps remain. Stir into the beef mixture and cook until thickened, about 1 minute more. Taste and adjust the seasoning, adding black pepper and more salt and/or curry powder, as desired. If you want more heat, stir in cayenne to taste. Stir in the peas. Refrigerate the filling until ready to use.

3. When you are ready to bake, preheat the oven to 400°F. Line a half sheet pan with parchment paper.

Recipe continues on next page

4. Take one log of dough and evenly divide the dough into six pieces. Roll out each piece of dough on a lightly floured surface into a round with a thickness of about ⅛ inch. Using a 6- to 7-inch saucer or template, cut out each round with a knife. Holding the dough round in your hand, brush the outer rim of each dough circle with the egg wash. Spoon a generous tablespoon of cooled filling into the center. Pull up the sides of the dough circles to meet on top of the filling. Pinch the edges together to form a crimped and fluted seal. Set the pasty, crimped seal up, on the prepared sheet pan. Repeat with the remaining filling and pastry.

5. When all the pasties are formed, cut slits into the sides to allow steam to escape during baking. Brush the remaining egg wash over the pasties.

6. Bake in the middle of the oven for 15 minutes. Decrease the oven temperature to 350°F and bake for 30 to 35 minutes longer, until the pasties are golden, rotating the sheets about halfway through the baking.

7. Don't be alarmed if the pasties have released fats and juices. Transfer the pasties to a paper towel–lined sheet pan and let cool for at least 10 minutes before serving.

LAMB AND FETA PASTIES

Pasties are distinguished from other handhold pies by their firm, strong dough with its distinctive crimp along the edge. Tallow makes the strongest dough that allows you to stuff in the filling and shape the pastry without tearing it. These pasties could be taken down into a mine — or into an office or to a picnic — because they transport well. They take a little time to make, but my family loves, loves, loves them.

Makes 10–12 pasties

1 tablespoon (0.45 ounce/13g) tallow or lard
1 pound ground lamb
1 leek, white and tender green parts, sliced
8 ounces thin-skinned potatoes (1 to 2 potatoes), peeled and julienned
8 ounces feta cheese
Salt and freshly ground black pepper
Pasty Pastry (page 273)
1 egg beaten with 1 tablespoon water, for egg wash

1. Heat a large skillet over medium-high heat. Add the tallow and let melt. Add the lamb and brown it, stirring, about 8 minutes. Add the leek and potatoes and sauté briefly until the leeks are wilted, about 2 minutes. Stir in the feta and season to taste with salt and pepper. Refrigerate until ready to use.

Recipe continues on next page

Lamb and Feta Pasties, continued

2. Preheat the oven to 400°F. Line a half sheet pan with parchment paper.

3. Evenly divide the dough into ten pieces. On a very lightly floured surface, roll out each piece of dough into a round with a thickness of about ⅛ inch. Using a 6-inch saucer or template, cut each round with a knife. Place a heaping ¼ cup of the filling mixture onto the center of the pastry round, leaving about a 1-inch margin of uncovered dough. Brush the outer rim of the dough circle with the egg wash. Pull up the sides of the dough circle to form a half circle. Pinch the edges together to form a good seal. To properly crimp the edge of the pastry, push down on the edge of the pasty and, using your index finger and thumb, twist the edge of the pastry over to form a crimp. Repeat this process along the edge of the pasty. When you've crimped along the edge, tuck the end corners underneath. Slightly flatten the bottom and set the pastry, crimped seal up, on the prepared sheet pan. Repeat with the remaining filling and pastry. You can gather together the pastry scraps to make more pasties if you have leftover filling.

4. When all the pasties are formed, cut a slit into the sides to allow steam to escape during baking. Brush the remaining egg wash over the pasties. (If you don't have enough egg wash remaining, add a little water to what remains to stretch it out.)

5. Bake for 15 minutes. Decrease the oven temperature to 350°F and bake for 30 to 35 minutes longer, until the pasties are golden.

6. Let cool for about 10 minutes before serving.

FRIED TACOS

Like many New Englanders, my first taste of Mexican food came at a Taco Bell, and it was kind of wonderful, discounting, of course, the abysmal working conditions of the fast-food workers and the farm workers who supply the chain, and the low quality of the ingredients. Mr. Bell (Glen Bell, the founder of Taco Bell) invented the mass-produced hard-shell taco that has had a grip on the American palate for far too long. The hard shell is brittle and breaks apart far too easily, hence the fried taco made from a soft corn tortilla and found at some pretty terrific taco trucks across the United States. Frying in lard or tallow is yet another step up for a crisp shell, without brittleness or grease. If you like, substitute your favorite taco filling for the one here.

Serves 4

2 tablespoons (0.8 ounce/25g) lard or tallow,
   plus 1 cup (7 ounces/200g) for frying
2 tablespoons chili powder, plus more as
   needed
1 tablespoon ground cumin
1 pound ground beef
1 onion, diced
3 garlic cloves, minced
1 cup beef or chicken broth
2 tablespoons masa harina mixed with
   3 tablespoons water
   Salt and freshly ground black pepper
2½ cups thinly sliced iceberg lettuce
½ cup chopped fresh cilantro
10 ounces Monterey Jack or cheddar cheese,
   grated
12–16 soft corn tortillas
   Salsa, guacamole, and/or sour cream, for
   serving

1. Melt the 2 tablespoons lard in a large skillet over medium-high heat. Add the chili powder and cumin and stir until blended with the fat. Add the beef, onion, and garlic, and cook, stirring, until the beef is browned, about 8 minutes. Stir in the broth and let cook for 5 minutes. Stir the masa-water mixture into the meat and continue to cook until the mixture is dry, not drippy, about 5 minutes. Season to taste with salt and pepper. Remove from the heat.

2. The cooking goes quickly so organize your work space. Mix together the lettuce and cilantro. You will need a half sheet pan lined with paper towels to hold the cooked tacos, with the lettuce and cheese waiting nearby.

3. Begin melting the remaining 1 cup fat in a large skillet and heat until about 375°F. Heat the tortillas in the microwave wrapped in a damp towel for about 30 seconds, until soft and pliable.

4. Place about 2 tablespoons of the meat filling in the center of a tortilla, fold in half, and slide into the hot fat; if needed, briefly hold closed with a pair of tongs or a spatula. Repeat so you are frying three or four tacos at a time, for 2 to 3 minutes per side. As the tacos are done, remove them with tongs, letting any excess fat drain out, and place on the lined sheet pan. Repeat with the remaining tacos.

5. While the subsequent batches of tacos cook, gently stuff each cooked one with a little cheese and a little lettuce; don't worry, they won't break.

6. When all the tacos are cooked and stuffed, serve, passing the salsa, guacamole, and/or sour cream at the table.

BEEF GOULASH

This stew (could be a soup) is known everywhere — except in Hungary — as goulash, made with a generous amount of paprika and cooked down until the meat is fork-tender. In Hungary, a similarly flavored soup is known as *gulyás* (from which the word "goulash" is derived), while a similar stew is known as *pörkölt*. In any case, recipes for goulash are found throughout Central Europe, Eastern Europe, the Netherlands, Belgium, Switzerland, Scandinavia, and Southern Europe. After reading dozens of recipes, I created my own recipe, which turns out to be quite similar to Polish iterations.

Serves 6

- 3 tablespoons unbleached all-purpose flour
- 1 teaspoon fine sea salt, plus more as needed
- 1 tablespoon dried thyme
- ¼ teaspoon freshly ground black pepper
- 2½ pounds beef chuck, cut into 1-inch cubes
- ½ cup (3.5 ounces/100g) any animal fat (lard is traditional)
- 2 onions, diced
- 2 red bell peppers, diced
- 2 Hungarian hot wax peppers, or similar mild to medium-hot chiles, stemmed, seeded, and diced
- 4 garlic cloves, minced
- ¼ cup Hungarian sweet or hot paprika
- 1 cup chicken broth
- ½ cup red wine
- 2 tablespoons tomato paste
- 1 tablespoon soy sauce
- 3 bay leaves
- 4 potatoes (about 1 pound), peeled and cubed
- 3 carrots, diced
  Freshly cooked egg noodles, for serving
  Sour cream, for serving

1. Preheat the oven to 275°F.

2. Combine the flour, salt, thyme, and black pepper in a large bowl. Add the meat and toss to coat.

3. Heat the fat in a large Dutch oven over medium-high heat. Add half of the beef and cook, turning, until browned, about 8 minutes. Transfer the meat to a bowl with a slotted spoon and brown the remaining beef. Transfer the meat to the bowl with a slotted spoon.

4. Add the onions to the pan and cook, stirring, until lightly caramelized, about 8 minutes. Stir in the bell and hot wax peppers and garlic, cook for 1 minute, then stir in the paprika.

5. Add the broth and stir to scrape up any browned bits stuck to the bottom of the pan. Return the beef to the pan and add the wine, tomato paste, soy sauce, and bay leaves. Bring to a boil, then cover and transfer to the oven.

6. Braise in the oven for 2½ to 3 hours, until the beef is tender. Add the potatoes and carrots and return to the oven to cook for 1 hour more, until the potatoes are tender.

7. Taste and add more salt if needed. Serve the stew on a bed of egg noodles, passing the sour cream at the table.

BEEF AND MUSHROOM PIE

Remember those little frozen pot pies? This pie is similar, only twenty times better, and it makes great use of leftovers — and whatever root vegetables you found in your CSA bag. If you happen to have any drippings or gravy from your roast, add it to the broth for even more flavor. I can't begin to describe how much my family enjoys this dish.

Serves 4–6

 Double-Crust Pastry made with lard or tallow (pages 270 and 272)
8 ounces thin-skinned potatoes, diced (peeling is optional)
 Salt
2 cups peeled and diced root vegetables (golden beets, carrots, celery root, parsnips, rutabagas, salsify, and/or turnips)
¼ cup (1.7 ounces/50g) any animal fat
1 large leek, white and tender green parts, sliced
12 ounces cremini or button mushrooms, sliced
⅓ cup unbleached all-purpose flour
2 cups beef broth
2½ cups diced roasted or braised beef
1 teaspoon dried thyme
 Freshly ground black pepper

1. Divide the pastry in half, with one half slightly larger than the other. Form each piece into a disk, wrap each in plastic wrap, and chill while you prepare the filling.

2. Preheat the oven to 425°F with a rack in the lower third of the oven.

3. Put the potatoes in a saucepan and add water to cover by 1 inch. Generously add salt to the water and bring to a boil. When the potatoes have boiled for 5 minutes, add the root vegetables. Continue to boil until the potatoes and root vegetables are just barely tender, about 5 minutes. Drain and set aside.

4. Melt the fat in a large saucepan over medium heat. Add the leek and mushrooms and sauté until the juices released by the mushrooms have mostly evaporated, about 10 minutes.

Recipe continues on next page

Beef and Mushroom Pie, continued

5. Add the flour and cook, stirring constantly, for 2 minutes. Stir in the broth, scraping up the browned bits, and bring the mixture to a simmer; simmer until the liquid thickens, about 5 minutes. Add the beef, thyme, and potatoes and root vegetables, stirring to combine. Season to taste with salt and pepper. Set aside while you roll out the pastry.

6. On a lightly floured surface, roll out the larger disk of pastry to a thickness of about ⅛ inch and large enough to overhang a 10-inch pie dish. Line the pie dish with the pastry. Spoon the filling on top. Roll out the remaining pastry and place on top of the filling. Seal and crimp the edges. Cut several vents in the crust to let steam escape.

7. Bake for 10 minutes. Reduce the oven temperature to 350°F and bake for 45 to 50 minutes longer, until the crust is golden brown and the filling is bubbling. Let it cool for at least 10 minutes before serving.

MANGALITSA PORK

My first bite of pork from a Mangalitsa pig was so extraordinary, so flavorful and juicy — and yes, so fatty — that I knew the two-hour trip I took to buy these chops from Eastman Farm in Barnard, Vermont, was worth it, and then some.

Mangalitsa (also known as Mangalica) pigs are a funny-looking breed, like what a cartoonist might sketch if a farmer crossed a sheep with a pig. They are covered with a curly hair that provides excellent insulation, making them the kind of pig that thrives in the outdoors. In fact, they cannot be raised in confinement, as more modern pork breeds can. Mangalitsas are excellent foragers and prefer the nuts and tubers they can root up in a forest. Their diet supplemented with wheat or barley, these slow-growing lard-type pigs develop meat that tastes like pork should — and definitely not "the other white meat." They reach maturity at around 15 months, compared to the 8 months of modern pig breeds.

The Mangalitsa breed was developed in the early 1800s from older hardy types of Hungarian pigs crossed with a Serbian breed and wild boar with the goal of producing a fast-growing lard hog that did not require any special care. The name of the breed translates from the Serbian as "hog with a lot of lard" because the carcass is 60 to 70 percent fat, compared with today's leaner hogs, which are less than 45 percent fat (Grade A).

In the days before refrigeration, the Mangalitsa was prized for its (shelf-stable) lard and the sausages made from its flesh, most typically seasoned with salt, pepper, and sweet Hungarian paprika, which gave the sausage a vibrant red color. The sausage was cold smoked and aged for two to three months, to create a meat that would keep for months and months. This type of sausage is still made today, and is especially prized when made with home-grown paprika. The Mangalitsa was also introduced into Spain, where its hams and loins are allowed to mature very slowly into a ham with exceptionally rich aromas and flavor, almost indistinguishable from the famed Ibérico ham.

The breed remained popular until the 1950s, when the demands of the new communist government meant that attention was turned to the farming of leaner animals that produced more meat more rapidly. Cheap cooking oils were promoted over lard for cooking. The Mangalitsa breed almost died out, but recent efforts in Hungary to revive its popularity have proved highly successful, and the pigs are now also being raised elsewhere in Europe and in the United States.

Researchers from the University of Auburn, in Alabama, have found that Mangalitsa meat is considerably higher in monounsaturated, or "good," fatty acids and lower in undesirable saturated fat than pork from Yorkshire hogs (another heritage breed).

If you want to sample this extraordinary meat, considered the "Kobe beef of pork" — and you should — you have to put effort into finding it, and you may have to order it online. When you do, be sure to consider getting some fatback to make Whipped Lardo (page 104); it will melt in your mouth. Many restaurants in Hungary will just serve slices of fatback as a dish on its own with just a sprinkling of sea salt to add taste and texture. Because of its high content of unsaturated fat, the slices will begin to melt as you place them on your tongue, releasing an incredible flavor across your mouth that some call pork candy.

Winter Squash with Duck Fat–Caramelized Apples, page 196

Side Dishes

REFRIED BEANS

Refried beans make a great side dish for any Mexican meal, but you can also use the beans as a filling for burritos or enchiladas. For more great piggy flavor, dice a small piece of salt pork and cook it with the beans. For a creamier texture (if you are serving the beans as a side dish), stir in 1 cup of grated Monterey Jack cheese.

Serves 6–8

- 2 cups dried pinto or black beans, soaked overnight and drained
- 8 cups water
- 2 onions, diced
- 2 garlic cloves, peeled
- 2 teaspoons fine sea salt, plus more as needed
- 1 teaspoon dried oregano
- 6 tablespoons (2.6 ounces/75g) any animal fat (lard is recommended)

1. In a large pot, cover the beans with the water. Add half of the onions, the garlic, salt, and oregano, and bring to a boil over high heat. Reduce the heat, partially cover (keep the pot lid off center), and simmer until the beans are very, very tender, about 1 hour. (I always sample at least 5 beans before calling the beans done.)

2. In a large skillet, heat the fat over medium-high heat until shimmering. Add the remaining onions and sauté, stirring occasionally, until translucent and lightly golden, about 7 minutes. Remove some beans from the pot and add to the skillet, being careful to avoid spatters. Using a bean masher, potato masher, or the back of a wooden spoon, smash the beans to form a chunky purée. Continue to add the beans and some cooking liquid until all the beans are added, along with about 1 cup of the bean cooking liquid. Reduce the heat to medium and cook, stirring, until the desired consistency is reached. If the refried beans become too dry, add more cooking liquid, 1 tablespoon at a time, as needed. Season with salt and serve.

BOSTON BAKED BEANS

A signature dish of New England, beans baked in an earthenware pot and sweetened with maple syrup was a recipe taught to early colonists by Native Americans. In the early 1800s, Boston was one point of the triangle trade involving slaves, rum, molasses, cod, wood, and other goods. Maple syrup gave way to molasses (except in Vermont), and salt pork enriched the beans. The beans can be served as a side dish or a topping for whole-grain toast as a simple supper dish.

2 cups dried small white beans, such as navy beans, soaked overnight and drained

1 onion, diced

12 cups water

2 teaspoons fine sea salt, plus more as needed

8 ounces salt pork or bacon, finely chopped

½ cup dark molasses (not blackstrap)

2 teaspoons Dijon or brown mustard

1 teaspoon ground allspice

Freshly ground black pepper

1. Combine the beans and onion in a large Dutch oven or ovenproof pot and add the water and salt. Bring to a boil, reduce the heat to a simmer, and cook, uncovered, until the beans are fully tender, about 45 minutes. Taste at least 5 beans from different spoonfuls to be sure they are fully tender before going to the next step.

2. Preheat the oven to 250°F or pull out a slow cooker.

3. Drain the beans, reserving all the cooking water. Return the beans to the Dutch oven or transfer to the slow cooker and add enough of the reserved cooking liquid to completely cover the beans; you will not use it all. Add the salt pork, molasses, mustard, allspice, and pepper to taste, and stir until well blended.

4. Transfer the Dutch oven to the oven and bake, uncovered, for about 4 hours, until the beans are extremely tender but still mostly whole, stirring a few times and adding more of the reserved cooking liquid (switching eventually to boiling water if you run out) as needed to prevent the beans on the surface from drying out. Stop adding liquid during the last hour of baking unless the beans look as though they are really drying out. Alternatively, cook in the slow cooker over high heat with the lid partially off for about 4 hours, stirring a few times, and adding reserved cooking liquid as needed.

5. Remove the beans from the oven or turn off the slow cooker and stir the beans carefully. The beans should be in a thickened glaze. If it's too dry, add boiling water sparingly until a glaze is achieved; if it's too wet, simmer briefly on the stovetop until reduced to the desired consistency. Adjust the seasoning with salt and pepper.

6. Keep the beans warm until you are ready to serve. Leftovers keep well. Reheat in a saucepan, adding water gradually as needed to loosen them up.

BACONY ROASTED BRUSSELS SPROUTS

Bacon and Brussels sprouts make a classic pairing. Here's a simple method for roasting them to perfection.

Serves 4

- 4 ounces bacon, diced
- 2 pounds Brussels sprouts
- Salt

1. Preheat the oven to 425°F.

2. Arrange the bacon in a single layer on a half sheet pan. Roast for about 8 minutes, until crispy.

3. Remove the bacon bits with a slotted spoon or pancake turner and drain on paper towels. Turn and tilt the pan until the fat evenly coats the pan. Add the Brussels sprouts and turn and toss with two silicone spatulas to coat the sprouts with the rendered fat. Spread out in a single layer.

4. Roast for about 15 minutes, until the sprouts are tender and browned. Transfer the sprouts to a serving bowl, sprinkle with the reserved bacon, season with salt, if needed, and serve hot.

ROASTED ROOT VEGETABLES

This is a jewel of a dish in appearance and flavor — especially if you take care to select several different colors of vegetables, cut them into even-size small (½-inch) pieces, and use a flavor-enhancing animal fat, such as duck fat. This dish needs no enhancement, but you could give it a light drizzle of pomegranate molasses or maple syrup, if you were so inclined.

Serves 4

- 3 tablespoons (1.4 ounces/40g) any animal fat (poultry fat is recommended)
- 6 cups peeled and diced mixed root vegetables (golden beets, carrots, celery root, parsnips, rutabagas, salsify, and/or turnips)
- 4 shallots, quartered
- Salt and freshly ground black pepper

1. Preheat the oven to 450°F with a rack in the center of the oven. Place a half sheet pan in the oven to preheat at the same time.

2. Add the fat to the hot sheet pan and turn and tilt the pan until the fat melts and covers the bottom of the pan. Add the vegetables and shallots, and toss with two silicone spatulas until evenly coated with fat. Spread out in a single layer and season with salt and pepper. Roast for about 30 minutes, until tender and lightly browned, turning the vegetables once about 20 minutes in. Serve hot.

NINE TIPS FOR BETTER ROASTED ROOT VEGETABLES

Whether your root cellar is filled or you're simply tempted by the colorful fall displays of root vegetables at the farmers' market, why not consider roasting those roots? It enhances their sweetness and mellows the sometimes sharp sulfurous notes of rutabagas and turnips.

Properly roasted roots are nicely browned on the outside and tender within; they are never pale in color or soggy in texture. Here are nine ways to guarantee delicious roasted root vegetables every time.

- Cut the vegetables into uniform small pieces. I prefer cubes, about ½ inch in size — never more than 1 inch. This creates the maximum surface area for caramelizing the sugars inherent in the roots. The number one mistake people make is cutting their vegetables into large pieces or chunks that cook unevenly.

- Mix it up. An assortment of root vegetables is more interesting than just one variety. Here's a dish where red beets can be mixed with other vegetables without turning the entire dish purple.

- Use a large sheet pan, two or more if you must. Sheet pans are preferable to any pan that has high sides. Never crowd the pan or the vegetables will steam rather than sear. Yes, the space demand does make it a problem when cooking for crowds, especially because the vegetables reduce so much in volume.

- Consider roasting with rendered poultry fat for great flavor. The fat will instantly melt on a preheated sheet pan. Lard or tallow adds less flavor but either is still a good choice, and bacon fat isn't out of the question.

- Flavor enhancers should include onions or shallots, which may burn but still taste sweet. Dried herbs and garlic can be added, but wait until the last 10 minutes of roasting; otherwise they will burn and become bitter.

- Roast in a hot oven and on the bottom rack. I generally roast at 425° to 450°F. If you are roasting more than one pan, use a convection oven if you can or consider roasting the sheet pans one at a time. If you must roast the veggies all at once, be sure to rotate the pans top to bottom as well as turning from side to side during roasting.

- Roast root vegetables for 25 to 30 minutes, until tender and well caramelized.

- If you are serving a crowd, consider roasting earlier in the day to avoid crowded pans, but reheat in the least crowded way possible. Or serve at room temperature on a bed of greens with a drizzle of salad dressing, like a maple-balsamic vinaigrette.

- Sprinkle with coarse salt before serving. For variety, consider drizzling with a maple syrup, boiled apple cider, or pomegranate molasses.

DUCK FAT–ROASTED BRUSSELS SPROUTS WITH POMEGRANATE MUSTARD DRIZZLE

Roasting is the very best way to prepare Brussels sprouts — with any animal fat. While many of the recipes gathered in this collection are traditional, this one is not. Do not overcrowd the pan or the lovely sprouts will steam rather than sear.

Serves 4

BRUSSELS SPROUTS

¼ cup (1.7 ounces/50g) duck fat (or any poultry fat

2 pounds Brussels sprouts, halved or quartered if large

2 apples

½ cup walnut pieces

2 garlic cloves, minced

DRIZZLE

2 teaspoons pomegranate molasses

2 teaspoons balsamic or apple cider vinegar

1 teaspoon Dijon mustard

1 teaspoon pure maple syrup

1. Preheat the oven to 425°F. Place a half sheet pan in the oven to preheat at the same time.

2. Put the duck fat on the hot sheet pan and turn and tilt until the fat evenly coats the pan. Add the Brussels sprouts and toss to coat with the duck fat. Spread out in a single layer.

3. Roast for about 10 minutes, until the sprouts are mostly tender and lightly colored. Add the apples, walnuts, and garlic, and continue roasting for 5 minutes longer, until the sprouts are lightly browned and tender, stirring or shaking the pan once for even cooking.

4. While the Brussels sprouts roast, prepare the drizzle. Combine the molasses, vinegar, mustard, and maple syrup in a small bowl and whisk to blend. Drizzle over the hot Brussels sprouts and apples and serve immediately.

ROASTED CABBAGE

Red, green, savoy — it's all good with this method of preparation. And the preferred fat? As with all roasted vegetables, poultry fat and bacon fat are particularly recommended, with bacon fat giving a whole different, and delicious, flavor profile. Tallow and lard work just fine, but contribute less flavor. The timing on this one depends on the age of the cabbage; the older the cabbage, the drier it will be, and the quicker the timing.

Recipe continues on next page

Serves 3 or 4

- 1 small head red or green cabbage
- 3 tablespoons (1.4 ounces/40g) any animal fat (poultry fat or bacon grease is recommended)
- Salt and freshly ground black pepper

1. Preheat the oven to 450°F. Place a half sheet pan in the oven to preheat at the same time.

2. Halve the cabbage, remove the core, and slice into 1- to 2-inch pieces.

3. Add the fat to the hot sheet pan and turn and tilt the pan until the fat melts and evenly coats the pan. Add the cabbage and toss with two silicone spatulas until the cabbage is evenly coated with the fat.

4. Roast for about 20 minutes, until the cabbage is browned in spots and tender. Season with salt and pepper and serve hot.

CRANBERRY–BRAISED RED CABBAGE

Red cabbage is often braised with apples, but sometimes I like to switch it up with cranberries, which add a lovely sweet-tart flavor to the cabbage. Use any cranberry sauce you have on hand, whether homemade or from a can, whole-berry or gelled.

Serves 6

- ¼ cup (1.7 ounces/50g) any animal fat
- 1 onion, halved and sliced
- 2 garlic cloves, minced
- 1½ teaspoons fine sea salt, plus more as needed
- ½ teaspoon freshly ground black pepper, plus more as needed
- ½ teaspoon ground allspice
- 1 red cabbage, halved, cored, and cut into 2-inch pieces
- ⅓ cup apple cider vinegar
- ½ cup cranberry sauce

1. Melt the fat in a large Dutch oven over medium heat and add the onion and garlic. Sauté until softened, about 3 minutes, then stir in the salt, pepper, and allspice, and cook for 1 minute.

2. Add the cabbage and toss until it is shiny and well coated. Add the vinegar, reduce the heat to low, stir well, cover, and cook for 20 minutes, stirring occasionally to ensure it doesn't stick.

3. Stir in the cranberry sauce and cook for 10 minutes longer. Season to taste with salt and pepper and serve.

BATTER-DIPPED AND CRUMBED DEEP-FRIED CAULIFLOWER

Snack or side dish? Either way, these bite-size morsels of vegetable goodness are irresistible. The coating holds up surprisingly well, though don't count on much in the way of leftovers.

Serves 4–6

- 1½ cups unbleached all-purpose flour
- ½ cup cornstarch
- ½ cup freshly grated Parmesan cheese
- 1 teaspoon baking powder
- 1 tablespoon garlic powder
- 1 tablespoon sweet or hot paprika
- 2 teaspoons dried Italian seasoning
- 2 teaspoons fine sea salt
- 1 teaspoon freshly ground black pepper
- 1 cup water
- Lard or tallow, for deep-frying
- 2 cups panko breadcrumbs
- 1 head cauliflower, cut into florets

1. Whisk together the flour, cornstarch, Parmesan, baking powder, garlic powder, paprika, Italian seasoning, salt, and pepper in a large bowl. Add the water and mix until the batter is smooth and thick.

2. Begin heating the fat to 365°F in a deep-fryer, wok, or large heavy Dutch oven. Line a large sheet pan with wire racks. Place the breadcrumbs in a shallow bowl.

3. Add the cauliflower pieces to the batter, turning to coat them evenly. A piece at a time, lift the cauliflower out of the batter and roll in the breadcrumbs until coated thoroughly. Place on a wire rack over the sheet pan. Repeat until all the cauliflower pieces are coated.

4. Working in batches, deep-fry the cauliflower pieces in the hot oil until golden brown, stirring to turn the pieces as needed. Lift the pieces out of the fat with a wire-mesh spider or tongs and allow to drain on the wire rack.

5. Serve immediately, or hold in a 200°F oven for up to 30 minutes. Serve warm.

ROASTED CAULIFLOWER STEAKS WITH ETHIOPIAN SPICE

A cauliflower steak is made by slicing the whole head vertically, including the stem. This gives the cauliflower a flat surface for more even browning, though you may only get three steaks and a lot of smaller pieces from one head. This recipe has been a family favorite for years, but it was a revelation when I made it with duck fat instead of olive oil. The browning was more even and the flavor was enhanced. If you can't find berbere spice (it is available online), substitute curry powder or another savory spice blend.

Serves 3 or 4

- 3 tablespoons (1.4 ounces/40g) any poultry fat, melted
- 1 teaspoon berbere (Ethiopian spice mix)
- 1 head cauliflower
  Salt

1. Preheat the oven to 425°F. Brush a half sheet pan or shallow roasting pan with some of the melted fat.

2. Stir together the berbere with the remaining fat.

3. To prepare the cauliflower, cut off any leaves and trim off the dried end of the stem. Slice the head into ½-inch thick slices, cutting through the core. Arrange the cauliflower slices in a single layer on the pan, flat side down. Brush the seasoned fat onto each cauliflower piece. Sprinkle with salt.

4. Roast for about 20 minutes, turning the cauliflower once halfway through, until browned and tender. Serve warm.

GARLIC- AND CHEESE-CRUMBED CAULIFLOWER

Here's a simple way to dress up a vegetable. The method can be adapted to other vegetables, such as green beans or broccoli. If your freezer is full of homegrown vegetables, you can blanch the frozen vegetables just long enough to heat them in step 2. If you're cooking something else and have the oven at a different temperature, just adjust the timing accordingly.

Serves 4

- 1 large head cauliflower, broken into florets
- ¼ cup (1.7 ounces/50g) any poultry fat or bacon grease
- 2 garlic cloves, minced
- 1 cup panko breadcrumbs
- 2 tablespoons minced fresh parsley
  Salt and freshly ground black pepper
- ½ cup freshly grated Parmesan or Gruyère cheese

1. Preheat the oven to 350°F.

2. Bring a large pot of salted water to a boil. Add the cauliflower and boil until tender, about 6 minutes. Drain, return to the pot, and cover to keep warm.

3. Melt the fat in a large ovenproof skillet over medium heat. Add the garlic and sauté just until fragrant, about 30 seconds. Add the breadcrumbs and parsley and toss until the crumbs are lightly toasted, about 4 minutes. Add the cauliflower and toss until the cauliflower is mixed into the crumbs. Add salt and pepper to taste.

4. Top the cauliflower with the Parmesan and bake until the cheese is melted, about 15 minutes. Serve hot.

SOUTHERN–STYLE GREENS

Collards are the toughest of the greens, but a Southern-style braise tenderizes the leaves and perfectly complements their green flavors. But don't stop there: Cabbage and kale are also perfect cooked this way.

Serves 4

- 4 ounces bacon or salt pork, diced
- 1 onion, diced
- 2 garlic cloves, minced
- 4 cups chicken broth
- 1 tablespoon Louisiana-style hot sauce, such as Frank's RedHot
  Salt and freshly ground black pepper
- 3 bunches collard greens, kale, mustard greens, or turnip greens, or 1 head green cabbage, chopped into 1-inch pieces

1. Cook the bacon in a large saucepan over medium heat until the fat renders and the bacon is crisp, about 8 minutes. Remove and reserve the bacon.

2. Add the onion and garlic to the fat in the saucepan and cook until the onion is limp and translucent, about 5 minutes. Add the broth and hot sauce and season to taste with salt and pepper. Bring to a boil.

3. Add the greens, reduce the heat to a simmer, cover, and cook until tender, about 1 hour for collards, 30 to 45 minutes for other greens.

4. Serve topped with the reserved bacon and pass more hot sauce at the table. The "pot liquor" can be reserved for soup — a frugal choice — or for moistening corn bread, as is traditional.

WILTED DANDELION GREENS

This classic salad is a favorite way to serve the bracingly bitter dandelion green. Foragers know to pick the greens in early spring, when the leaves are just emerging, and before the flowers appear. That's when they are the most tender and least bitter. Once the flower appears, the greens become tough and more bitter, although still edible. Cultivated varieties are significantly larger, and the tougher ends of the stems should be trimmed away. Cooking these greens with bacon cuts some of the bitterness, while the vinegar cuts the richness of the bacon. Use the highest-quality vinegar you have on hand; it makes a difference.

Serves 4

    4 slices thick-cut bacon
    1 shallot, diced
    2 tablespoons apple cider vinegar (preferably
        unfiltered artisan cider; see Note)
    2 tablespoons pure maple syrup
    1 bunch (about 1 pound) dandelion greens,
        chopped
       Salt and freshly ground black pepper

1. Cook the bacon in a large skillet over
    medium heat until the fat renders and
    the bacon is crisp, 8 minutes. Remove the
    bacon strips from the skillet and drain on
    paper towels.

2. Return the skillet to medium heat to reheat
    the bacon grease. Add the shallot and sauté
    until translucent, about 3 minutes. Stir in
    the vinegar and maple syrup and heat for
    1 minute. Add the dandelion greens and
    sauté just until wilted, about 2 minutes.
    Season to taste with salt, if needed, and
    pepper. Arrange on four warmed serving
    plates, crumble a slice of bacon over each
    serving, and serve immediately.

    NOTE: If you don't have a high-quality
    apple cider vinegar, it is best to substitute
    a high-quality wine vinegar or balsamic
    vinegar.

SOUTHERN-STYLE GREEN BEANS

Having grown up on lightly steamed green beans,
these slow-cooked green beans were a revelation
of flavor. And if you are a lazy gardener who lets
your green beans get large before harvesting,
you'll be happy to know this recipe is perfect for
green beans of any size or degree of tenderness.

Serves 4

    4 ounces bacon or salt pork, diced
  1½ pounds green beans, ends trimmed
    1 onion, halved and sliced
    4 cups chicken broth
   ½ teaspoon freshly ground black pepper
   ½ teaspoon red pepper flakes
       Salt

1. Brown the bacon in a large saucepan until
    crisp, about 8 minutes. Remove the bacon
    and set aside on paper towels to drain.

2. Add the green beans to the saucepan along
    with the onion, broth, black pepper, and
    pepper flakes. Add salt to taste. Bring to
    a boil, then reduce the heat to low, cover,
    and simmer for about 1½ hours, stirring
    occasionally.

3. To serve, drain the beans and transfer to a
    serving bowl. Sprinkle the bacon on top.
    Serve hot. Reserve the pot liquor for soup
    or corn bread.

SPAGHETTI SQUASH WITH BACON

Because bacon makes everything taste better, this is a recipe for those who don't really care for winter squash — or so says one of my sons. I think winter squash is terrific in general, but spaghetti squash is uniquely bland and needs a strong flavor boost, which this recipe provides.

Serves 4

> 1 (4-pound) spaghetti squash
> 4 ounces bacon, diced
> 4 garlic cloves, minced
> 2 tablespoons granulated or brown sugar (optional)
> Salt and freshly ground black pepper

1. Preheat the oven to 375°F.

2. Slice the spaghetti squash in half lengthwise. Scrape the seeds and fibers from the center of the squash and discard (or save the seeds for roasting and eating as a snack). Place the squash cut side down in a baking dish and add about 1 inch of water to the dish. Bake for about 1 hour, or until a sharp knife can be easily inserted with only a little resistance.

3. Meanwhile, fry the bacon in a large skillet over medium heat until crisp, about 8 minutes. Add the garlic and sauté for 30 seconds, until fragrant.

4. When the squash is baked through, use a fork to scrape the flesh from the squash in long strings into the skillet. Toss the squash to mix in with the bacon and to coat with the bacon grease. Add the sugar, if desired (much depends on the squash itself, and the cure on the bacon). Add salt and pepper to taste and serve hot.

WINTER SQUASH WITH DUCK FAT–CARAMELIZED APPLES

I can't imagine anyone not enjoying this appealing combination of apples and squash. Generously seasoning with salt and pepper is essential. Because it is such a crowd-pleaser, and because it can be made in advance and reheated, it makes a perfect side dish for the Thanksgiving table.

Serves 4–6

> 1 large buttercup, butternut, or red kuri squash, or ½ small blue Hubbard squash
> Duck Fat–Caramelized Apples (page 252)
> ¼ teaspoon freshly grated nutmeg
> Salt and freshly ground black pepper

1. Preheat the oven to 400°F.

2. Cut the squash into halves if small, or into quarters if large. Scrape out the seeds and fibers from the squash and discard (or save the seeds for roasting and eating as a snack). Place the squash skin side up

in a baking dish and add about 1 inch of water to the dish. Bake for 1 to 1½ hours, depending on the size of the pieces, until completely tender when pierced with a fork.

3. When the squash is done, drain off the water. Turn the pieces flesh side up and allow to cool enough to be easily handled. Scrape the flesh from the skins into a large bowl and discard the skins. Mash or beat until very smooth.

4. Fold in the apple mixture and the nutmeg. Season to taste generously with salt and pepper.

5. If desired, reheat in the microwave or in the top of a double boiler set over boiling water. Serve hot.

ROSEMARY ROASTED POTATOES

Okay, I understand that nobody wants to succumb to heart disease. But when the medical authorities blamed it on animal fats, how did the population agree to give up fat-basted roasted potatoes? These potatoes are golden on the outside, creamy on the inside. Roasting with any animal fat will yield a perfect crust, but poultry fat gives it the best flavor of all.

Serves 4

¼ cup (1.7 ounces/50g) any animal fat
2 garlic cloves, minced
1 teaspoon finely chopped fresh or dried rosemary
3 pounds thin-skinned potatoes, cut into 2- to 3-inch chunks
Salt and freshly ground black pepper

1. Preheat the oven to 450°F. Place a half sheet pan in the oven to preheat at the same time.

2. Combine the fat, garlic, and rosemary in a small glass or ceramic bowl. Microwave on high for 1 minute, then let sit while the oven preheats (or let sit for at least 10 minutes, if the oven is already preheated).

3. Remove the hot sheet pan from the oven and strain the fat onto the pan, reserving the garlic and rosemary. Turn and tilt the pan until evenly coated with the fat. Add the potatoes and toss with two silicone spatulas to coat with the fat. Sprinkle with salt and pepper.

4. Roast for about 20 minutes, until the potatoes will release from the pan when you try to turn them with a metal spatula. Turn the potatoes and continue roasting for 20 minutes longer, until tender and golden brown throughout. Sprinkle with the reserved garlic and rosemary and serve hot.

POTATO LATKES

The Jewish holiday of Hanukkah falls sometime in December. It is traditional to eat fried foods then, commemorating the lamp oil in the temple in Jerusalem that burned for eight days when supplies should have lasted only for a night. Fried potato pancakes, or latkes, are probably the most famous of the traditional foods. The perfect latke is crisp on the outside, tender and snowy white on the inside; it is never greasy or gray in color. There are a few extra steps in my recipe as part of my never-ending quest to get this dish right. Most important is the frying medium. It wasn't until I switched from a vegetable seed oil to a poultry fat — as my ancestors used — that the pancakes stopped being greasy.

Serves 4–6

- 3 pounds russet or other baking potatoes, peeled
- 1 large onion

  Fresh lemon juice or distilled white vinegar, as needed
- 2 eggs, lightly beaten
- 2 teaspoons fine sea salt
- ¼ teaspoon freshly ground black pepper
- ¼ cup (1.7 ounces/50g) any poultry fat (goose fat is traditional)

  Applesauce, for serving

  Sour cream, for serving

1. Coarsely grate the potatoes and onion with a box grater or in a food processor.

2. Transfer the potato mixture to a large bowl filled with acidulated water (1 tablespoon lemon juice or vinegar to 4 cups water). Swish around with your hands for 1 minute. Pour into a strainer and drain well. Place a clean kitchen towel on the counter. Dump the potatoes onto the towel and pat dry. This step will keep the potatoes from turning pink, then gray, as they are exposed to air.

3. In the food processor, pulse half of the potato mixture until finely chopped but not puréed. Combine the grated and finely chopped potatoes in a large bowl and add the eggs, salt, and pepper. Mix well.

4. Preheat the oven to 200°F. Line a half sheet pan with wire racks.

5. Heat the fat in a large skillet over medium-high heat. (If your skillet is less than 12 inches in diameter, start with half the poultry fat and add more as needed as you make more batches.) Working in batches, drop the potato mixture into the hot fat, ¼ cup at a time, and fry until golden on the bottom, 2 to 3 minutes. Turn and fry on the other side, 2 to 3 minutes. Place the latkes on the prepared sheet pan and put the sheet pan in the oven to keep warm. Continue making latkes until all the batter is used.

6. Serve as soon as possible. Pass the applesauce and sour cream at the table.

GERMAN POTATO SALAD

In the category of everything tastes better with bacon, this is one of the original dishes that proves the concept. Although called a salad because a vinaigrette is made with the bacon grease and vinegar, it is really a warm side dish. And it isn't really German; it is a classic French preparation, but gets the name "German" because it has bacon. If you can, use an artisanal unfiltered cider vinegar; it adds great flavor to this simple dish.

Serves 4–6

- 2 pounds waxy, thin-skinned potatoes, halved and sliced ¼ inch thick
- 1 tablespoon fine sea salt, plus more as needed
- 8 ounces bacon, diced
- 1 shallot or small onion, diced
- 5 tablespoons apple cider vinegar, plus more as needed
- 1 tablespoon whole-grain mustard, plus more as needed
- 1 tablespoon brown sugar, plus more as needed
  Freshly ground black pepper
- ¼ cup chopped fresh parsley

1. Put the potatoes and salt in a large saucepan and cover with cold water. Bring to a boil, cover, reduce the heat, and simmer until the potatoes are tender, 5 to 10 minutes. Reserve ¼ cup of the cooking liquid and drain the potatoes. Transfer the potatoes to a large bowl and cover to keep warm.

2. Meanwhile, cook the bacon in a medium skillet over medium heat until brown and crisp, about 8 minutes. Remove the bacon with a slotted spoon and transfer to the bowl with the potatoes.

3. Add the shallot to the skillet and cook until slightly softened, about 3 minutes. Stir in the reserved cooking liquid, vinegar, mustard, and brown sugar. Bring to a boil. Pour the mixture over the potatoes and toss to coat with a silicone spatula. Taste and add salt, if needed (the bacon may be salty), and pepper. Add additional vinegar, mustard, and brown sugar if the salad needs extra punch. Add the parsley and toss lightly. Serve immediately.

POTATOES ANNA

I've had my heart broken by this dish many times because it is as beautiful as it is exquisitely delicious when it releases from the pan properly. But it can stick to the pan if the pan isn't the exact right pan, and therein lies the heartbreak if you have spent the time trying to make it beautiful for a special dinner. So follow my lead and line your pan with parchment paper. I don't think there are enough superlatives in the English language to adequately describe this dish.

Serves 4–6

- ¼ cup (1.7 ounces/50g) any poultry fat
- 3½ pounds Yukon Gold or baking potatoes, peeled and thinly sliced on a mandoline to a ¹⁄₁₆-inch thickness
- 1½ teaspoons fine sea salt
- ¼ teaspoon freshly ground black pepper

1. Preheat the oven to 450°F with a rack in the middle of the oven.

2. Melt the fat in a 10-inch cast-iron skillet over low heat. Remove the skillet from the heat and pour the melted fat into a large bowl. Add the potatoes, sprinkle with the salt and pepper, and toss gently with a silicone spatula to coat.

3. Line the bottom of the skillet with a round of parchment paper. Arrange the potatoes in overlapping concentric circles in the skillet. You should be able to make at least three layers. If there is any fat remaining in the bowl, scrape it onto the potatoes. Press down on the potatoes with a wide metal spatula.

4. Bake the potatoes for a total of 45 minutes, pressing down on the potatoes with a wide metal spatula after 20 minutes. The potatoes are done when the outside edge is golden brown and the potatoes in the center are tender when pierced with a fork. Let stand, covered, at room temperature for 5 minutes.

5. Carefully loosen the edges with a heat-proof flexible spatula. Invert a plate with a rim over the skillet. Using pot holders and holding the plate and skillet together firmly, invert the skillet. The potatoes should drop onto the plate. Peel off the parchment paper to reveal a beautiful dish of lightly browned, concentric potato slices. Slice into wedges to serve.

POTATOES RÖSTI

Add onions and this rösti becomes hash browns, a pedestrian dish that is well served by cooking in any animal fat. But rösti, which is found throughout Sweden, is simple and sophisticated, especially if made with goose fat or duck fat. It may take some practice to get the flipping technique perfected, but it is well worth it.

Serves 4

1¼ pounds thin-skinned potatoes
    Salt and freshly ground black pepper
3 tablespoons (1.4 ounces/40g) any animal fat
    (duck fat or goose fat is recommended)

1. In a saucepan, cover the potatoes with water. Generously salt the water. Bring to a boil and parboil the potatoes until just barely tender, but not soft, about 7 minutes. Allow to cool in the water for 5 minutes; then drain and chill for at least 2 hours, or overnight.

2. Coarsely grate the potatoes and season with salt and pepper. Heat 2 tablespoons of the fat in a 10-inch cast-iron skillet over medium heat, then add the grated potato. Allow to cook for a couple of minutes, and then shape it into a flat cake, pressing down lightly with a wide metal spatula. Continue to cook for a couple of minutes, then gently shake the pan to loosen the potato. Cook until golden and crisp on the bottom, about 10 minutes.

3. To turn the cake, place a plate on top of the pan and invert it so the cake sits cooked side up on the plate. Add the remaining 1 tablespoon fat to the pan and, when hot, slide the potato cake back into the pan, keeping it cooked side up. Cook until the bottom is golden, 5 to 10 minutes. Serve hot.

DUCK FAT OVEN FRIES

Let's face it: Poultry fat is just too precious to use in a deep fryer because the quantities of fat required are too great. That's where oven roasting comes in. Made with any poultry fat, these "fries" are so good, you won't even want to dip them in ketchup.

Serves 4

4 large russet or other baking potatoes, cut
    into ⅓-inch-wide sticks
½ cup (3.5 ounces/100g) duck fat (or any
    poultry fat)
    Coarse sea salt

1. Put the potatoes in a large bowl and cover with cold water. Let stand for at least 20 minutes, or up to 6 hours.

2. Preheat the oven to 425°F. Place a half sheet pan in the oven to preheat at the same time.

3. Drain the potatoes in a colander and lay them out on thick towels to dry.

4. Put the duck fat on the hot sheet pan and tilt and turn until the fat evenly coats the pan. Add the potatoes and toss with two silicone spatulas until the potatoes are well coated with the melted fat.

5. Roast for 20 to 25 minutes, turning once for even cooking, until the potatoes are tender when pierced with a fork and evenly browned. Serve hot.

HAND-CUT FRIES

French fries at home are tricky. They require advance planning and prepping, really good knife work, and time to work in small batches that need to be fried twice. But if you are going to the trouble of making French fries at home, do make them with tallow — the fat that made McDonald's fries so successful.

Serves 4

> 2 pounds russet or other baking potatoes
> Tallow, for deep-frying
> Coarse sea salt or kosher salt

1. Cut a thin slice along the length of one side of a potato to provide a flat surface. Turn it on its cut side and slice the potato lengthwise into ½-inch-thick planks. Cut the planks lengthwise into ½-inch-wide sticks. It is important to have all the fries even in size. Transfer the potato sticks to a bowl of cold water and swish them around to release the starch on the surface of the potatoes. Rinse, add clean water, and swish around again. Repeat until the water is clear. Spread the fries on a clean towel and pat dry.

2. Begin heating the tallow in a heavy saucepan or deep fryer to 325°F. Fry in batches until the edges just start to color, 3 to 5 minutes. (This step cooks them through.) Remove from the hot fat with a wire-mesh spider into a wire basket or onto a wire rack. Let cool for at least 30 minutes — better still, let freeze overnight. If you like, the partially cooked frozen potatoes will keep for up to 2 months.

3. When you are ready to serve, reheat the fat to 375°F. Fry in batches until golden brown and crispy, 3 to 4 minutes. Transfer out of the fat with a spider into the wire basket or onto the rack. Drain briefly, then spread out on a paper towel–lined sheet pan, pat dry with more paper towels, and season with salt. Serve immediately.

HOME FRIES

Home fries used to be a standard item in diners in the Northeast, but they were often disappointing, greasy, not sufficiently browned. Then diner cooks started deep-frying their home fries, solving the problem of greasiness and insufficient browning. But those cubes of potatoes are French fries in disguise, not home fries. Animal fats allow for superior browning, superior flavor, and no greasiness — just lovely potatoes browned with onions and the perfect accompaniment to your eggs.

Serves 4

- 2 pounds thin-skinned or new potatoes, cut into bite-size cubes (peeling is optional)
  Fine sea salt
- 5 tablespoons (2.3 ounces/65g) any animal fat
- 1 onion, halved vertically and sliced into thin slivers
  Freshly ground black pepper

1. Put the potatoes in a saucepan, cover with cold water, and salt generously. Bring to a boil over high heat and boil until the potatoes are just tender, but not mushy, about 7 minutes. Drain the potatoes in a colander but do not wash the saucepan.

2. Return the saucepan to the stove, add 2 tablespoons of the fat, and melt over medium heat. Add the potatoes to the saucepan and toss to coat in the fat.

3. Melt another 2 tablespoons of the fat in a large skillet over medium-high heat and heat the fat until shimmering. Add the potatoes, reduce the heat to medium, and cook until crisp and golden brown on all sides, about 10 minutes, turning occasionally with a spatula. Do not turn the potatoes until they are ready to release from the pan. Return the potatoes to the saucepan.

4. Add the remaining 1 tablespoon fat to the skillet and heat until shimmering. Add the onion and cook, turning occasionally, until tender and lightly browned, about 5 minutes. Return the potatoes to the skillet with the onion, season with salt and pepper, and cook until heated through and browned some more, about 10 minutes. Serve immediately.

POULTRY FAT FRIES

Many bars and bistros are serving duck fat fries and chicken fat fries. These are potatoes that have been fried in poultry fats. How can they afford to do it?

Usually these establishments do a first fry in (inexpensive) vegetable oil to blanch the potatoes. The second fry, done to order, is in poultry fat. It gives the fries a touch of poultry flavor, which translates on the palate as a delightful richness.

HASSELBACK POTATOES

Created and named in a Stockholm restaurant, these dressed-up roasted potatoes are sometimes called hedgehog potatoes because of their appearance. The goal is a potato that is crispy on the outside and creamy inside, like a roasted potato, and looks terrific.

Serves 4

- 4 russet or other baking potatoes
- 4 thyme sprigs
- ¼ cup (1.7 ounces/50g) any animal fat (tallow is recommended), melted
- ¼ cup panko breadcrumbs
- 3 tablespoons freshly grated Parmesan cheese
- 2 garlic cloves, minced
- 1 teaspoon fine sea salt

1. Preheat the oven to 450°F. Grease a shallow baking dish just large enough to hold the potatoes without crowding.

2. Peel the potatoes. Place on a cutting board and cut a narrow slice from the bottom of each potato. (This helps the potato lie flat and not roll; the slices can be dipped in the melted fat and roasted with the potatoes. The slices will be done after 45 minutes — cook's treat.) Place a wooden chopstick on each side of the potato lengthwise. Use a sharp knife to slice each potato crosswise, making the cuts ¼ inch apart, cutting down vertically. The chopsticks will prevent the knife from cutting entirely through the potato.

3. Chop the thyme sprigs and tuck in between the potato slices. Put the potatoes in the baking dish, cut side up. Generously brush the potatoes with the melted fat and bake for 45 minutes.

4. While the potatoes bake, combine the breadcrumbs, Parmesan, garlic, and salt in a small bowl and mix together.

5. When the potatoes have baked for 45 minutes, remove from the oven and brush the potatoes with the remaining melted fat (remelt as needed). Pat the breadcrumb mixture on top of each potato and drizzle any remaining melted fat on top.

6. Return the potatoes to the oven and bake for an additional 15 to 30 minutes, until the potatoes are crispy on the outside and tender within and the crumbs are golden. Serve hot.

POTATO TAQUITOS

Taquitos are great for making ahead and reheating for breakfast, grab-and-go snacks, or serving as a side dish or main dish. With a filling of mashed potatoes, they become the Mexican equivalent of Jewish knishes.

Makes 16 taquitos

- 1½ pounds potatoes, peeled and cut into chunks
- 2 garlic cloves, peeled
- 1 tablespoon fine sea salt
- ½ cup (3.5 ounces/100g) lard
- 2 scallions, white and green parts, minced
- 2 tablespoons finely chopped fresh cilantro
- 1 or 2 jalapeño, Anaheim, or other fresh chiles, minced (seeding is optional), or 2 tablespoons minced roasted green chiles
- ½ teaspoon ground cumin
- 16 corn tortillas
-   Salsa, for serving
-   Chopped lettuce, for serving
-   Sliced avocado or guacamole, for serving
- 2 limes, cut into wedges, for serving

1. Put the potatoes and garlic in a saucepan, cover with water, add the salt, and bring to a boil over high heat. Boil until the potatoes are completely tender, about 20 minutes. Reserve ½ cup of the potato water. Drain the potatoes in a colander, then dump into a large bowl.

2. Add 2 tablespoons of the lard and mash, mixing in the reserved potato water to give the potatoes the texture of mashed potatoes. Fold in the scallions, cilantro, chiles, and cumin.

3. Place the tortillas between two damp paper towels. Microwave for 1 minute, or until warm and pliable.

4. Spoon about 2 tablespoons of the potato filling onto a tortilla and form into a sausage shape. Tightly roll up into a tube and place on a sheet pan, seam side down. Repeat until all the filling is used.

5. Melt the remaining 6 tablespoons lard in a large cast-iron skillet over medium heat. Place the taquitos, as many as will fit in one layer, seam side down in the skillet. Slowly brown on all sides, turning every other minute or so. This might take up to 10 minutes. Transfer to a paper towel–lined plate and keep warm. Repeat with the remaining taquitos.

6. Serve warm, passing the salsa, lettuce, avocado, and limes at the table.

SPICED SWEET POTATO OVEN FRIES

The devil is in the details when it comes to making sweet potato oven fries. The potatoes must be cut to a uniform thickness or they will cook unevenly. The pan must not be too crowded or the potatoes will steam and never brown, but the pan must not be too empty or the potatoes will burn. The semolina flour helps the mix of spices cling and aids in browning. The mix of spices is, however, just a suggestion and can be varied at will. The sweet potato fries will not become as crisp as oven fries made with white potatoes, nor will they brown uniformly. But they are entirely delicious.

Serves 2–4

- 2 large sweet potatoes, peeled and cut into ¼-inch sticks
- 1 tablespoon semolina flour
- 1 tablespoon smoked paprika
- 1 teaspoon fine sea salt
- 1 teaspoon onion powder
- 1 teaspoon garlic powder
- ¼ teaspoon freshly ground black pepper
- ¼ teaspoon ground allspice
- ¼ cup (1.7 ounces/50g) any animal fat
  Coarse sea salt or kosher salt (optional)

1. Preheat the oven to 400°F. Place a half sheet pan in the oven to preheat at the same time.

2. Soak the potato sticks in cold water to cover for 30 minutes. Drain and dry on a layer of towels.

3. Combine the semolina, paprika, fine salt, onion powder, garlic powder, pepper, and allspice in a large bowl and mix well. Add the sweet potatoes and toss to coat with two silicone spatulas.

4. Remove the hot sheet pan from the oven and add the fat. Tip and turn the pan until it is well coated and the fat is completely melted. Add the sweet potatoes, use two silicone spatulas to toss to coat with the fat, and place the pan on a rack in the lower third of the oven. Roast for 30 to 35 minutes, until the potatoes are well colored, turning once or twice as the potatoes brown on the bottom (check before turning), and shaking the pan for even cooking. (The timing will depend on how crowded the pan is.)

5. Taste and sprinkle with coarse salt, if desired. Serve hot.

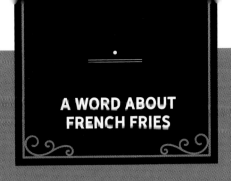

A WORD ABOUT FRENCH FRIES

McDonald's rise to popularity may have had as much to do with the perfection of their French fries as with the cheapness of the burgers and the speed by which they are produced. Originally those fries were cooked in 93 percent beef tallow and 7 percent cottonseed oil, until customer demand for less saturated fat prompted a switch to hydrogenated vegetable oil in 1990. (Such was our understanding of fat and health at the time.)

When trans fats came to be understood as a health threat, McDonald's and other fast-food chains switched to nonhydrogenated vegetable oils. This, in turn, unleashed a host of new problems — the heated oxidized oils form polymers that collect as a swamp of gunk at the bottom of the fryer. Airborne particles of these same polymers cause a buildup on all the restaurant surfaces, coating everything with a sticky shellac. It's quite possible that workers' lungs are placed at risk from these toxic particles, according to the World Health Organization.

McDonald's fries are also given a bath of dextrose (sugar) to guarantee even browning. In order to replicate the flavor that beef tallow conferred, McDonald's adds "natural beef flavor," which contains hydrolyzed wheat and milk (in case you thought they were gluten-free and dairy-free).

The fries are fried briefly at the factory, frozen, then shipped to the individual franchises, where the fries are given a second fry. The second fry, like the first one, is in a blend of canola oil, soybean oil, and hydrogenated soybean oil, plus corn oil and an additive called TBHQ, or tertiary butylhydroquinone, to reduce oxidation. Unfortunately, at high doses TBHQ has been shown to induce stomach tumors in rats.

I'll stick to making my French fries at home in beef tallow.

LITHUANIAN POTATO PUDDING

I grew up eating potato kugel — a pudding of grated potato and onions, enhanced with chicken schmaltz. It's a dish that is best served with drippings from a roast because it is rather plain. The Lithuanian version (*kugelis*) is enhanced with bacon and makes a delicious dish that could be served as a main dish or a side dish; it makes a great brunch dish. Lithuania (for the geographically challenged) is the southernmost of Europe's Baltic states, a former Soviet bloc nation bordering Poland, Latvia, and Belarus. In that area of the world, potato puddings are common, with many variations.

Serves 6–10

- 3 pounds russet or other baking potatoes, peeled
- 2 onions, peeled and quartered
- 12 ounces bacon, diced
- 4 eggs
- 1 cup milk
- 1 teaspoon fine sea salt
- 1 teaspoon freshly ground black pepper
- 3 tablespoons (1.4 ounces/40g) lard

1. Preheat the oven to 350°F. Place a 9- by 13-inch baking dish in the oven to preheat at the same time.

2. Shred the potatoes in a food processor or with the coarse holes of a box grater. Transfer to a bowl and fill with cold water. Swish the potatoes around to release the starch. Pour off the starchy water, then run the potatoes under cold running water until the water runs clear. Transfer the potatoes to a colander to drain.

3. Shred the onions in the food processor or grate with the box grater and add to the potatoes.

4. Cook the bacon in a large skillet over medium heat until the bacon has softened and begun to release its fat, about 5 minutes. Remove from the heat.

5. In a large bowl, combine the potatoes, onions, and bacon plus any bacon grease. Beat the eggs and milk in a bowl until well combined and stir into the potato mixture. Stir in the salt and pepper.

6. Remove the preheated baking dish from the oven. Add the lard and turn the baking dish to coat well. Add the potato mixture, spreading it out evenly.

7. Bake for about 1¼ hours, or until golden on top. Remove and let stand for 15 minutes before slicing into squares. Serve hot.

KASHA VARNISHKAS

This perfect combination of nutty buckwheat and pasta, lightly flavored with onion, is a dish of my youth, with one slight variation. Toasted buckwheat groats are found whole in the bulk grain section of most natural food stores or ground and packaged in small boxes in supermarkets that have a section of kosher foods. By using the whole grain (really a seed), you avoid the necessity of toasting with an egg white to avoid clumping — a step that is recommended on the box. Traditionally, this Jewish dish is made with chicken or goose fat, but any animal fat can be used. I've tested it with tallow, and it was almost as good as when it is made with a poultry fat. I just can't bear to test it with lard (too many ancestors spinning in their graves), but I'm sure it would be fine.

Serves 4–6

- 1½ cups small bow tie pasta
- 3 tablespoons (1.4 ounces/40g) any animal fat (poultry fat is recommended)
- 1 large onion, diced
- 1 cup roasted buckwheat groats
- 2 cups chicken broth

  Salt and freshly ground black pepper

1. Bring a pot of salted water to a boil and cook the pasta according to the package directions. Drain and set aside.

2. Heat the fat in a large heavy skillet over medium-high heat. Add the onion and sauté until the onion is transparent, about 3 minutes. Add the buckwheat and sauté until the groats are well toasted, 3 to 5 minutes. Slowly add the broth, then add salt to taste. Reduce the heat, cover, and cook over very low heat, without stirring, until the liquid has been absorbed and the grains are tender, 15 to 20 minutes.

3. Remove from the heat and stir in the pasta. Wipe the pot lid dry and cover the pot, placing a crumpled clean kitchen cotton or paper towels between the lid and pot so excess moisture is absorbed. Let stand for 5 minutes. Then fluff with a fork, season with more salt, if needed, and pepper and serve.

Duck Fat–Caramelized
Apples, page 252

BAKED GOODS AND DESSERTS

POPOVERS

In this country, popovers are made with butter and served with jam. But in England, popovers are most often made with beef drippings from a roast and called Yorkshire pudding. Call these delightful puffs of eggy bread whatever you like, serve them however you like, and explore the subtle differences you'll taste if you make these with either tallow skimmed from roast beef or a poultry fat.

Makes 12 popovers

- 1 cup unbleached all-purpose flour
- 1 teaspoon fine sea salt
- 4 eggs
- ¾ cup whole milk
- ½ cup (3.5 ounces/100g) poultry fat or tallow, melted

1. Beat together the flour, salt, eggs, and milk in a bowl until smooth. Let the batter rest for at least 30 minutes, or up to overnight. Refrigerate if the batter is to sit for more than 30 minutes, but bring to room temperature before baking.

2. Spoon 2 teaspoons of the fat into each cup in a 12-cup muffin pan. Preheat the oven to 425°F. Place the muffin pan set on a rack in the center of the oven to preheat.

3. Pull the hot muffin pan out of the oven and place on a heatproof surface. Fill each cup about two-thirds full with the batter. Immediately return to the oven and bake for 20 minutes, then reduce the oven temperature to 350°F and continue baking for 20 to 25 minutes, until the popovers are puffed and deeply browned. The longer you bake, the more set the interior, so use the shorter times if you want the interior to be custardy.

4. Remove from the pan and serve hot.

BUTTERMILK BISCUITS

Because lard and tallow have higher melting temperatures than butter, they tend to make taller biscuits, especially when you use the folding, or laminating, technique I use below. These biscuits are crispy on the outside and tender on the inside, perfect for enjoying with chicken pot pie filling (page 156) or the old Southern favorite, sausage gravy (page 166).

Makes 9–12 biscuits

- 3 cups unbleached all-purpose flour
- 2 tablespoons baking powder
- 1 tablespoon sugar
- 1½ teaspoons fine sea salt
- ⅔ cup (4.4 oz/130g) any animal fat (except bacon grease)
- 1 cup buttermilk

1. Preheat the oven to 450°F with a rack in the middle of the oven. For ease of cleanup, line a sheet pan with parchment paper or a silicone baking mat.

2. Combine the flour, baking powder, sugar, and salt in a bowl and whisk to combine.

3. If using lard or poultry fat, cut in the fat with a pastry cutter or rub in with your fingertips until the mixture resembles coarse crumbs. If using tallow, combine the tallow and ½ cup of the flour mixture in a food processor fitted with a steel blade and process until it has the consistency of tiny pebbles. Add the rest of the flour mixture and pulse to mix in. Return to the bowl.

4. Pour in the buttermilk and stir until all the flour is moistened. Using your hands, gather the dough together into a ball; you may have to knead the dough a bit to incorporate all the flour, but handle the dough as little as possible.

5. Turn the dough out onto a lightly floured board and pat into a rectangle. Fold in the sides, as though you were folding a piece of paper into thirds. Pat out again. Turn the dough 90 degrees and fold in the sides again, and then pat out. Do this a few times, to create layers in the dough; this process is called laminating. Pat out the dough to a thickness of about ½ inch. Cut into 2½-inch rounds. Place the biscuits on the prepared sheet pan about 1 inch apart. Gather the scraps together, pat out, and cut all into rounds.

6. Bake for 15 to 18 minutes, until golden brown. Serve as soon after baking as possible. Biscuits are best on the day they are made, but day-old biscuits are delicious toasted.

CHEESE BISCUITS

These biscuits turn a simple meal of soup into a cause for celebration. My kids love biscuits in every iteration, but these may be their favorite. Cheese biscuits are delicious served with Bacon Jam (page 112).

Makes about 12 biscuits

- 3 cups unbleached all-purpose flour
- 2 tablespoons baking powder
- 1 tablespoon sugar
- 2 teaspoons fine sea salt
- 2 teaspoons dried Italian herbs
- 2 teaspoons garlic powder
- 2 teaspoons onion powder
- 1 cup grated cheddar cheese
- 6 tablespoons (2.6 ounces/75g) any animal fat (except bacon grease)
- 1½ cups buttermilk

1. Preheat the oven to 425°F with a rack in the middle of the oven. For ease of cleanup, line a sheet pan with parchment paper or a silicone baking mat.

2. Combine the flour, baking powder, sugar, salt, Italian herbs, garlic powder, onion powder, and cheese in a large bowl and mix well.

3. If using lard or poultry fat, cut it in with a pastry cutter or rub it in with your fingertips until the mixture resembles coarse sand with some larger pieces. If using tallow, combine the tallow with ½ cup of the flour mixture and process in a food processor fitted with the steel blade until the mixture has the consistency of tiny pebbles. Add the remaining flour mixture and pulse to mix in. Return the mixture to the bowl.

4. Stir in the buttermilk and knead briefly in the bowl to form a ball, making sure all the flour is incorporated. Handle the dough as little as possible to retain tenderness.

5. Turn the dough out onto a floured surface and pat it out into a rectangle. Fold in the sides as though you were folding a piece of paper into thirds. Turn the dough 90 degrees, pat out into a rectangle, and fold in the sides again. Repeat at least twice more. This process is called laminating, and it creates layers in the biscuit. Finally, pat out to a thickness of about ½ inch and cut the dough into 2½-inch rounds. Place the biscuits on the prepared sheet pan about 1 inch apart. Gather the scraps together, pat out, and cut all into rounds.

6. Bake for about 22 minutes, until golden brown. Serve as soon after baking as possible. Biscuits are best on the day they are made. Day-old biscuits, however, are delicious toasted.

EVERYDAY FLATBREADS

This recipe makes flour tortillas, but to call them tortillas limits their purpose. Yes, they make a great burrito or wrap for a fajita, but these flatbreads are equally good used as a wrap for everything from hummus to Indian lentils to lamb burgers to chicken salad. You can also take them hot off the pan, wrap them around thin slices of Italian cold cuts and cheese, and call them *piadina*.

Makes 8 flatbreads

- 2 cups unbleached all-purpose flour, plus more as needed
- 1 teaspoon fine sea salt
- ¼ cup (1.7 ounces/50g) lard
  Scant ⅔ cup warm (98° to 100°F) water

1. Mix the flour and salt in a food processor fitted with the steel blade. Add the lard and process until completely mixed in. Add the water and continue processing until the dough forms a ball. If the dough is tacky, knead in a little more flour, a tablespoon at a time (and add as little as possible). Cover and let rest on the counter for at least 30 minutes and up to 1 hour.

2. Divide the dough into eight equal pieces. Reshape each piece into a ball and roll out to form a thin round — as thin and as round and as evenly as you can. Stack between pieces of waxed paper or parchment paper.

3. Heat a dry griddle or large cast-iron skillet over high heat until hot. Cook the flatbreads one at a time for about 30 seconds on each side. There should be some dark brown (not burnt!) spots on each side to indicate when it is done.

4. Stack on a plate as you continue to cook them all. These are best if used immediately. If necessary, you can store them, wrapped in paper towels in a resealable plastic bag in the refrigerator; reheat them for a few seconds in a skillet before serving.

BUTTERMILK CORN BREAD

You could write a manifesto about American foods fueled by this corn bread — it tastes that elemental. This is an unrefined, unrepentant version, full of porky goodness, yet more corn than pork, more bread than meat. Use your "porky" lard if you want. Skip the cracklings if you must.

Serves 8

- 1 cup coarse yellow cornmeal
- 1 cup unbleached all-purpose flour
- 1 tablespoon sugar
- 1½ teaspoons fine sea salt (see Note)
- 1 teaspoon baking soda
- 1 teaspoon baking powder
- 1 egg, lightly beaten

1½ cups buttermilk

5 tablespoons (2.3 ounces/65g) lard

1 cup pork cracklings (optional)

1. Preheat the oven to 450°F. Place a 10-inch cast-iron skillet in the oven to preheat at the same time.

2. Whisk together the cornmeal, flour, sugar, salt, baking soda, and baking powder in a large bowl.

3. Beat the egg with the buttermilk in another bowl. Stir into the cornmeal mixture until smooth.

4. Remove the skillet from the oven and add the lard. Tip and swirl the pan until the lard melts and coats the pan. Pour most of the lard back out of the skillet and into the cornmeal batter, leaving enough to coat the pan. Add the cracklings to the batter, if desired, and stir until well blended. Pour the batter into the pan; it should sizzle vigorously. Shake the skillet to distribute the batter evenly.

5. Bake for 15 to 18 minutes, until a cake tester inserted into the center comes out clean. Serve warm, straight out of the skillet.

   NOTE: If the cracklings are salted, add 1 teaspoon salt instead of 1½ teaspoons.

CLASSIC WHITE SANDWICH BREAD

This is the classic all-American white sandwich loaf, upon which many depend for their better-than-average PB&J, BLT, or any other combination of ingredients one can imagine. This is an easy loaf to make. If you don't make bread regularly, you may want to make a double batch. Otherwise, there may not be bread left over for your morning toast, and it does make excellent toast.

Makes 1 loaf

1 cup milk

¼ cup (1.7 ounces/50 grams) lard or other animal fat (except bacon grease)

1 tablespoon honey

4 cups white bread flour, plus more as needed

1 tablespoon instant or quick-rise yeast

1 tablespoon fine sea salt

¼–½ cup water

1. Combine the milk, lard, and honey in a small saucepan over medium heat and heat just until the lard melts; you do not have to scald the milk. Alternatively, heat in a microwave just until the lard melts.

2. Combine the flour, yeast, and salt in a food processor fitted with the dough blade or in a stand mixer fitted with the dough hook attachment. With the machine running, add the milk mixture and process until well

Recipe continues on next page

combined. Scrape down the sides of the bowl and continue mixing, adding the water as needed to make a dough.

3. Turn the dough out onto a lightly floured work surface and knead until the dough feels smooth and elastic, about 5 minutes.

4. Place the dough in a lightly greased bowl, cover, and let it rise for 1 to 1½ hours, until the dough is puffy and almost doubled in size.

5. Grease a 5- by 9-inch loaf pan. Turn the dough out onto the work surface and pat gently to deflate. Roll the dough into a 9-inch log and place it seam side down in the loaf pan.

6. Cover the pan and let the dough rise for 1 to 1½ hours, until it has risen 1 to 1½ inches above the rim of the pan. Begin checking after 1 hour, and begin preheating the oven to 350°F shortly before the dough appears fully risen.

7. Bake the bread for 40 to 45 minutes, until the bread is golden brown and an instant-read thermometer inserted into the center reads 195° to 200°F. Turn the bread out onto a rack to cool. When completely cool, slice and serve. Wrap any leftover bread in plastic and store at room temperature.

AMISH POTATO BUNS

You couldn't ask for better soft American dinner rolls to serve on Thanksgiving. These are perfect for mopping up gravy as you finish your meal and for making snack-size sandwiches with the leftovers. There is a fairly long rising time, but the actual work is brief. The combination of fat and mashed potatoes in the dough keeps leftover rolls fresh for the next day.

Makes 12–16 buns

 2 large eggs
 ⅓ cup sugar
 6 tablespoons (2.6 ounces/75g) lard
 1 cup lightly packed mashed potatoes
2–3 teaspoons fine sea salt
2½ teaspoons instant or quick-rise yeast
 ¾ cup lukewarm water or milk
 4 cups unbleached all-purpose flour, plus more for kneading
 1 tablespoon butter, melted

1. Combine the eggs, sugar, lard, and mashed potatoes in a food processor or the bowl of a stand mixer fitted with the paddle attachment. If the mashed potatoes are seasoned leftovers, add 2 teaspoons salt. If the potatoes are unseasoned, add 3 teaspoons. Mix briefly, just to combine. Add the yeast, water, and flour, and mix together until you have a soft dough.

2. Turn the dough out onto a lightly floured work surface. The dough will be sticky. Flour your hands and knead briefly until you have a smooth, soft dough.

3. Place the dough in a lightly greased bowl, cover the bowl with plastic wrap, and let the dough rise until it's doubled in bulk, about 90 minutes.

4. Lightly grease a 9- by 13-inch baking dish. Turn the dough out onto a very lightly floured work surface. Divide it into 12 to 16 portions. Form each portion into a smooth ball, about the size of a Ping-Pong ball. Place the rolls in the prepared pan. Cover the pan with lightly greased plastic wrap.

5. Let the rolls rise for 1½ to 2 hours, until they're quite puffy. Preheat the oven to 350°F.

6. Bake the rolls for 20 to 25 minutes, until they're golden brown and the internal temperature reaches at least 190°F on an instant-read thermometer. Remove the rolls from the oven and turn them out of the pan onto a wire rack. Brush with the melted butter. Serve the rolls warm or at room temperature. Store the rolls, well wrapped in plastic, for several days at room temperature; freeze for longer storage.

NOTE: To make this dough with tallow instead of lard, weigh it out, then process it in the food processor with ½ cup flour. Then add the eggs, sugar, mashed potato, and salt and mix together. Proceed with the recipe as directed.

LITHUANIAN BACON ROLLS

Lasineciai, or Lithuanian bacon buns, are practically the national dish of Lithuania, a Baltic state in central Europe. The buns are made with a soft, brioche-like dough, stuffed with onion and bacon. These are best straight out of the oven, but they are perfectly fine served reheated or at room temperature. Every coffee shop in America should serve these tasty treats for a savory grab-and-go.

Makes 12–15 buns

DOUGH

3½ cups unbleached all-purpose flour, plus more as needed

2¼ teaspoons (1 packet) instant or quick-rise yeast

½ cup sugar

2 teaspoons fine sea salt

4 eggs

¼ cup water

½ cup (3.5 ounces/100g) lard, at room temperature

Recipe continues on next page

Lithuanian Bacon Rolls, continued

FILLING

  6 ounces bacon, finely chopped

  1 onion, finely chopped

  ½ cup water

  1 egg yolk beaten with 1 tablespoon water, for egg wash

1. To make the dough, combine the flour, yeast, sugar, and salt in the bowl of a stand mixer fitted with the paddle attachment. Mix briefly to blend. Combine the eggs, water, and lard in a bowl and beat until smooth. Add to the flour mixture and beat on medium speed until well blended. Replace the paddle with a dough hook and knead for at least 5 minutes, until the dough is smooth and shiny. Scrape the dough from the dough hook, cover the bowl, and let it sit for 1 to 3 hours, until doubled in bulk.

2. Meanwhile, make the filling. Heat a large skillet over medium heat. Add the bacon, onion, and water, and cook, stirring frequently, until the water has evaporated, the bacon is fully cooked (but not crisp), and the onion is cooked, about 20 minutes. Set aside.

3. Line a half sheet pan with parchment paper. Turn the dough out onto a lightly floured surface and gently fold it over several times until you have a smooth dough. It will still be soft and sticky.

4. Divide the dough into 12 to 15 even-size pieces. Stretch each piece of dough into a circle about 4 inches across. Place a scant tablespoon of the filling on each dough round. Bring up the sides of the dough to encase the filling. Firmly pinch the edges to completely seal the buns. Gently pull the bun into a torpedo shape and place seam side down on the prepared sheet pan. Repeat until all the dough and filling is used. Cover the buns and let them rise until puffy, 1 to 1½ hours.

5. Preheat the oven to 375°F. Brush the buns with the egg wash and bake for about 25 minutes, until the rolls are golden brown and the centers register 190°F on an instant-read thermometer.

BACON GREASE AND BAKED STUFF

I love bacon as much as the next baker, but I do not like bacon grease in my baked goods. The operative word here is "grease." Baked goods made with bacon grease instead of butter or lard taste greasy. Cooked bacon bits add an intriguing flavor to cookies and muffins and such, but save the bacon grease for frying your eggs or making grilled cheese sandwiches.

POLVORONES

Although Americans are most likely to have eaten these melt-in-your-mouth cookies as Mexican wedding cakes made with butter, this recipe originated in Spain. And at one point during the Spanish Inquisition, it was illegal to make these cookies with anything but lard (to smoke out Jews and Muslims, who are prohibited to eat pork by their religions). If that history weirds you out, by all means, call these cookies by one of their other American names, including snowballs, moldy mice, sandies, sand tarts, or butterballs. Or call them *biscochitos* (as they do in Mexico), tea cakes (as they do in Sweden and Russia), *dandulas kiflik* (as in Bulgaria), *biscochos* (as in Cuba), *des kourabi* (as in Greece), or *rohlichky* (as in Ukraine). A cookie so widely adopted, and so thoroughly time-tested, must be good, and it is. It is!

Makes about 42 cookies

- 1 cup pecans, walnuts, or hazelnuts
- 2 cups confectioners' sugar
- 1 cup (7 ounces/200g) lard, at room temperature
- ½ teaspoon pure vanilla extract
- 2 cups unbleached all-purpose flour
- ¾ teaspoon fine sea salt

1. Combine the nuts and 1 cup of the confectioners' sugar in a food processor. Process until the nuts are finely ground. Add the lard and vanilla and process until light and fluffy. Add the flour and salt and process until combined. Cover and refrigerate the dough for about 1 hour, or until firm.

2. Preheat the oven to 350°F. Line three large cookie sheets with parchment paper.

3. Form the dough into 1-inch balls and place about 2 inches apart on the prepared cookie sheets.

4. Bake, one cookie sheet at a time, for about 15 minutes, until the cookies look dry and cracked on top. The color will not change much.

5. While the cookies are baking, place the remaining 1 cup confectioners' sugar in a shallow bowl.

6. Cool the cookies on the parchment sheets for a few minutes. While the cookies are still warm, roll them in the confectioners' sugar. Place on a wire rack to finish cooling.

7. When the cookies have cooled completely, roll them again in the confectioners' sugar to give them an even coating. Store in an airtight container between sheets of parchment or waxed paper. They will keep well for at least 1 week.

JAM THUMBPRINT COOKIES

Call them thumbprints, thimble cookies, or Polish tea cookies. In Sweden, they are called *hallongrottor*, which means "raspberry caves." With so many names from so many cultures and so many variations, lard seems as likely as butter in the dough. In any case, this is a delicious, but not terribly sweet, cookie. They are especially beautiful if you fill some with raspberry jam and some with apricot or peach jam.

Makes about 60 cookies

2¼	cups unbleached all-purpose flour
1	cup confectioners' sugar, plus more for dusting
1	teaspoon fine sea salt
½	teaspoon baking soda
½	cup finely ground almonds
1	(8-ounce) package cream cheese, softened
¾	cup (5.5 ounces/155g) lard, at room temperature
½	teaspoon pure vanilla extract
⅓–½	cup jam of your choice

1. Sift together the flour, confectioners' sugar, salt, and baking soda. Stir in the ground almonds.

2. Combine the cream cheese and lard in a food processor and process until smooth. Add the vanilla and process until mixed in. Add the dry ingredients and process until well blended.

3. Form the dough into a ball and refrigerate for at least 30 minutes.

4. Preheat the oven to 350°F. Line three large cookie sheets with parchment paper.

5. Shape the dough into 1-inch balls. Place 1½ inches apart on the prepared cookie sheets. Using a sewing thimble or the handle of a wooden spoon, make a generous indentation in the center of each cookie, twisting the tool to spread the opening. (You can use your thumb, but a thimble or wooden spoon makes a neater hole.) Using the tip of a pointed teaspoon (not a measuring spoon), fill each hole with ¼ to ½ teaspoon jam. You want a generous amount of jam, but don't overfill; make the holes bigger if needed.

6. Bake, one cookie sheet at a time, for 14 to 16 minutes, until light golden brown. Cool on wire racks. Dust lightly with confectioners' sugar before serving. Stored in an airtight container, they will keep well for at least 1 week.

CUBAN SUGAR COOKIES

Torticas de Morón are shortbread cookies from the city of Morón, in central Cuba. These traditional cookies have dozens of variations, but are typically made with lard and a hint of lime or lemon.

Makes about 42 cookies

2⅓ cups unbleached all-purpose flour
     Finely grated zest and juice of 1 lime
  1 teaspoon fine sea salt
  1 cup sugar, plus more for sprinkling
  1 cup (7 ounces/200g) lard
  2 eggs, 1 whole and 1 separated
  1 teaspoon pure vanilla extract or rum extract

1. Whisk together the flour, lime zest, and salt.

2. Combine the sugar and lard in a food processor and process until well combined. Add the whole egg and the egg yolk and process until well blended. Add the vanilla and two-thirds of the lime juice (about 2 teaspoons) to the dough and process until blended. Add about one-third of the flour mixture and process until incorporated. Continue to add the flour mixture, a few tablespoons at a time, and process until all the flour is well blended into the dough. If you squeeze a pinch of dough, it should hold together.

3. Dump the dough onto a large sheet of waxed paper or parchment paper. Divide the dough in half and gather each portion of dough into a ball. With your hands, work the dough to form two cylinders about 1½ inches in diameter. Use the paper to roll the cylinders into smooth, evenly shaped logs. Roll the logs in the paper and chill for at least 30 minutes, or up to 2 days. (If chilling for more than a few hours, wrap the paper-covered cylinders in plastic.)

4. Preheat the oven to 325°F. Line three large cookie sheets with parchment paper.

5. Unwrap the dough cylinders and slice into ⅓-inch-thick cookies. Place the slices on the prepared cookie sheets about 1 inch apart; the cookies will spread out. Beat the egg white with the remaining lime juice (about 1 teaspoon). Brush the tops of the cookies with the egg wash and finish with a sprinkling of sugar.

6. Bake, one cookie sheet at a time, for about 25 minutes, until the cookies are very lightly touched with color around the edges and bottoms; they will not brown much. Transfer to wire racks to cool. Stored in an airtight container, they will keep well for at least 1 week.

OATMEAL—CHOCOLATE CHIP COOKIES

This is a supersize cookie, good for bake sales and other occasions when a cookie competes on the table with other goodies.

Makes 24–30 cookies

- 1½ cups unbleached all-purpose flour
- 1 teaspoon baking powder
- 1 teaspoon baking soda
- 1 teaspoon fine sea salt
- 1 teaspoon ground cinnamon
- ¾ cup (5.5 ounces/155g) any animal fat (except bacon grease)
- 1½ cups firmly packed light brown sugar
- 2 eggs, plus 1 egg yolk
- 1½ teaspoons pure vanilla extract
- 2 cups old-fashioned rolled oats
- 2 cups chocolate chips

1. Preheat the oven to 375°F. Line three large cookie sheets with parchment paper.

2. Whisk together the flour, baking powder, baking soda, salt, and cinnamon in a large bowl.

3. Combine the fat and brown sugar in a food processor. Process until well blended. Add the eggs, egg yolk, and vanilla, and process until smooth.

4. Add the flour mixture to the food processor and process until blended. Return the dough to the bowl and mix in the oats and chocolate chips.

5. For each cookie, drop one ice cream scoop of dough (about 2 rounded tablespoons) onto the parchment paper, placing them about 3 inches apart to allow the cookies to spread.

6. Bake, one cookie sheet at a time, for 11 to 13 minutes, until the cookies are light brown at the edges and dry to the touch but still soft in the middle. Let the cookies cool on the pan for a few minutes, then slip the parchment paper with the cookies still on them onto a counter to cool and set completely. Stored in an airtight container, they will keep well for at least 1 week.

CRINKLE–TOP MOLASSES COOKIES

Molasses was a common sweetener at a time when lard was the standard fat in kitchens throughout New England and the American South. Dipping the cookie dough in granulated white sugar gives the top its distinctive crackled and crunchy top.

Makes about 36 cookies

- ¾ cup (5.5 ounces/155g) any animal fat (except bacon grease)
- 1 cup firmly packed light or dark brown sugar
- ¼ cup molasses
- 1 egg
- 2¼ cups unbleached all-purpose flour
- 2 teaspoons baking soda
- 1 teaspoon ground cinnamon
- ½ teaspoon ground allspice
- ½ teaspoon fine sea salt
- 3 tablespoons granulated sugar

1. Preheat the oven to 325°F. Line three large cookie sheets with parchment paper.

2. Combine the fat and brown sugar in a food processor and process until smooth. Add the molasses and egg and process until well blended. Scrape down the sides of the bowl, then add the flour, baking soda, cinnamon, allspice, and salt. Process until the dough forms a ball.

3. Spoon the granulated sugar onto a small plate. Shape the dough into 1½-inch balls. Dip the tops into the granulated sugar and place, sugared side up, on the cookie sheets about 2 inches apart.

4. Bake, one cookie sheet at a time, for 13 to 16 minutes, just until set and the cookies appear dry. Slide the parchment paper off the cookie sheets to allow the cookies to cool before handling them. Stored in an airtight container, they will keep well for at least 1 week.

JELLY DOUGHNUTS

A delicious jelly doughnut is a cause for celebration, no? Jews enjoy jelly doughnuts on Hanukkah, fried in goose fat or schmaltz, and Germans eat Berliners (also a jelly doughnut), fried in lard, on New Year's Eve for good luck. Jelly doughnuts are easier to make than you might think, but it does take time because it is a yeast dough that needs time to rise and they are fried. It is essential that you get the temperature of the frying medium right — too hot and the outside gets too dark while the inside conceals a center of uncooked dough; too low and the doughnut is leaden.

Makes 15 doughnuts

2¾ cups unbleached all-purpose flour, plus more as needed

  1 tablespoon instant or quick-rise yeast

¼ cup sugar, plus about ½ cup for coating

¾ cup milk

  2 eggs

¼ cup (1.7 ounce/50g) poultry fat or lard, at room temperature, plus fat for deep-frying

½ teaspoon fine sea salt

About ¾ cup jelly or jam of your choice (jam should be fairly smooth)

1. Combine 2½ cups of the flour and the yeast in a bowl.

2. Combine the ¼ cup sugar, milk, eggs, fat, and salt in the bowl of a stand mixer fitted with the paddle attachment. Mix on medium speed until smooth. Turn the speed to low, and slowly add the flour mixture. Scrape down the sides of the bowl with a silicone spatula and switch to the dough hook attachment. With the machine running on low to medium-low speed, knead the dough until it is shiny, very smooth, soft but not mushy, and noticeably elastic, 6 to 7 minutes. This is a tacky, slack dough, and it will wrap around the dough hook and clear the side of the mixer bowl, but it will stick to the bottom, so add the remaining ¼ cup flour as needed. Alternatively, knead by hand on a lightly floured surface for about 5 minutes, until the dough is smooth and springs back to the touch.

3. Scrape the dough into a greased bowl, cover with plastic wrap, and let rise in a warm place until doubled in size, 2 to 3 hours.

4. Scoop the dough out onto a floured surface, punch it down, and roll it out to a thickness of ½ inch with a flour-dusted rolling pin.

5. Line a sheet pan with parchment paper and lightly sprinkle with flour.

6. Using a 2½-inch biscuit cutter, cut out the doughnuts and place on the prepared sheet pan a few inches apart. Handling the cut doughnuts without stretching them can be tricky, but if they are slightly misshapen, re-form them into rounds on the sheet pan. Gather together the scraps and roll and cut more doughnuts or form "doughnut holes" until all the dough is used. Cover with a clean kitchen towel and allow to rise for about 30 minutes, or until doubled again.

7. Begin heating fat in a deep fryer or saucepan for deep-frying and heat to 355° to 360°F. Line a sheet pan with wire racks. Place the ½ cup sugar in a shallow bowl.

8. When the doughnuts have risen and the fat is hot, gently lower some doughnuts into the fat with a wire spider, adding no more than three doughnuts at a time. Fry them until puffed and golden brown, about 3 minutes on each side, turning with the spider. Take the fried doughnuts out of the hot fat with the spider and place them on the lined sheet pan to drain.

9. While the doughnuts are still warm, dredge them in the sugar to coat and replace on the wire racks.

10. Fit a piping bag with a plain round ¼-inch tip and fill with the jelly. Poke a hole in the side of each doughnut with a chopstick and wiggle the chopstick to make a pocket. Squeeze about 1 tablespoon of jelly into each doughnut. Replace on the racks. Sprinkle some of the remaining sugar on top of each doughnut. These are best enjoyed on the day they are made, but placed in a paper bag, they hold up about as well as bakery doughnuts for enjoying on the second day.

BLUEBERRY MUFFINS

I have 12 highbush blueberry bushes, and the berries are large and plump, perfect for pies, ice cream, and toppings. But I also live near a wild blueberry area, where the berries are sweet and small — perfect for muffins because they distribute themselves so nicely and evenly in a batter. If you can find wild ones, use them here. Fresh berries are preferred to frozen ones, but if you are using frozen berries, adjust the baking time accordingly, and don't thaw first or your muffins will be a lurid purple in color.

This is a proof-of-concept recipe — proof that animal fats work fine in a muffin recipe. If you are used to coffee-shop muffins, which these days tend to be oversized and somewhat oily, these old-fashioned muffins will be a revelation. Feel free to substitute another fruit — raspberries or very finely chopped apples come to mind.

Recipe continues on next page

Blueberry Muffins, continued

Makes 12 muffins

- 2 cups unbleached all-purpose flour
- 2 teaspoons baking powder
- 1 teaspoon fine sea salt
- ½ cup (3.5 ounces/100g) any animal fat (except bacon grease)
- 1 cup granulated sugar
- 2 eggs
- 1½ teaspoons pure vanilla extract
- ½ cup buttermilk
- 2¼ cups fresh blueberries
- Turbinado, demerara, or other raw sugar, for sprinkling

1. Preheat the oven to 375°F. Line a 12-cup muffin pan with paper liners. If your pan is not nonstick, lightly grease the pan where the muffins will rise above the papers.

2. Whisk together the flour, baking powder, and salt in a bowl.

3. If you are using lard or a poultry fat, beat the fat and granulated sugar in the bowl of a stand mixer fitted with the whip attachment on medium speed until light and fluffy. Add the eggs one at a time, scraping down the sides of the bowl and beating well after each addition. Beat in the vanilla. If you are using tallow, combine the tallow and ½ cup of the flour mixture in a food processor and process until the mixture looks like tiny pebbles. Add the granulated sugar and process until well combined. Add the eggs and vanilla and process until well blended. Transfer to the bowl with the remaining flour mixture.

4. Fold in the flour mixture by hand with a rubber spatula, alternating with the buttermilk, until just combined. Add the berries and fold gently until evenly distributed. Do not overmix.

5. Scoop the batter into the prepared muffin pan (an ice cream scoop works well); the cups will be very full. Sprinkle the raw sugar evenly on top of the muffins.

6. Bake for about 30 minutes, until lightly golden and a cake tester comes out clean. Let the muffins cool in the pan for about 10 minutes. Run a knife around the edge of each muffin to free it from the pan if necessary, then transfer the muffins to a wire rack to cool. The muffins are spectacular while still warm, but they hold up nicely a second day.

BANANA–NUT BREAD WITH CHOCOLATE CHIPS

Banana bread's popularity rides in waves. Sometimes it seems like it is featured at every bake sale and brunch. Sometimes it just sort of falls off the radar. This version isn't likely to be forgotten — and it can be used to make over-the-top French toast or bread pudding.

Serves 8–10

- 1½ cups unbleached all-purpose flour
- 1½ teaspoons fine sea salt
- 1½ teaspoons baking powder
- 1½ teaspoons baking soda
- 1 teaspoon ground cinnamon
- ¼ teaspoon freshly grated nutmeg
- ½ cup (3.5 ounces/100g) any animal fat (except bacon grease)
- 1 cup firmly packed light or dark brown sugar
- 1 teaspoon pure vanilla extract
- 2 eggs
- 1 cup mashed ripe bananas (about 3 medium or 2 large bananas)
- ¼ cup plain yogurt
- 1 cup chopped toasted walnuts, almonds, or pecans
- 1 cup mini chocolate chips

1. Preheat the oven to 350°F. Lightly grease a 5- by 9-inch loaf pan and line with parchment paper so the ends of the paper overhang the edges of the pan.

2. Whisk together the flour, salt, baking powder, baking soda, cinnamon, and nutmeg in a medium bowl until well combined.

3. If you are using lard or poultry fat, combine the fat, brown sugar, and vanilla in a food processor and process until well combined. Add the eggs, mashed bananas, and yogurt, and again process until well combined. If you are using tallow, combine the tallow and ½ cup of the flour mixture in a food processor and process until the mixture looks like tiny pebbles. Add the brown sugar and vanilla and process until well combined. Add the eggs, mashed bananas, and yogurt and again process until well combined.

4. Add the banana mixture to the flour mixture, along with the nuts and chocolate chips, and fold together until all the flour is moistened. Spoon the batter into the prepared loaf pan, smoothing the top.

5. Bake the bread for 1 hour, then lightly lay a piece of aluminum foil over the top so it does not burn, and continue baking for 15 to 20 minutes longer, until a cake tester inserted into the center comes out mostly clean, with no uncooked batter.

6. Allow the bread to cool for 10 minutes in the pan. Remove it from the pan by lifting it out with the parchment paper, and let it cool completely on a wire rack.

PEACH COBBLER

High summer means peaches — juicy, sweet, and sun kissed. Under a blanket of not-very-sweet biscuit dough and accented with nutmeg, peaches make a perfect dessert. Serve straight from the oven and add a scoop of vanilla ice cream on top, enjoying the contrast of hot and cold, savory and sweet.

Serves 6

PEACH FILLING

6 or 7	ripe yellow peaches, (about 2½ pounds)
6	tablespoons sugar, or as needed
2	tablespoons cornstarch
½	teaspoon freshly grated nutmeg
1	tablespoon lemon juice

BISCUIT DOUGH TOPPING

1½	cups unbleached all-purpose flour
¼	cup sugar, plus 1 tablespoon for sprinkling
1½	teaspoons baking powder
1	teaspoon fine sea salt
6	tablespoons (2.6 ounces/75g) any animal fat (except bacon grease)
5	tablespoons milk

1. Grease an 8-inch square baking dish or a 1½-quart casserole dish.

2. To make the peach filling, bring a large pot of water to a boil and blanch the peaches for 30 to 60 seconds. Remove with a slotted spoon and immerse them in a bowl of cold water. The skins will then slip off easily. Remove the skins, cut each peach in half, and discard the pits. Cut the peaches into ½-inch slices. Transfer to the prepared baking dish.

3. Combine the sugar, cornstarch, and nutmeg in a small bowl and mix well. Sprinkle over the peaches. Drizzle on the lemon juice. Toss the peaches, mixing gently until they are thoroughly coated. If the peaches are particularly tart, add more sugar to taste.

4. Set the filling aside for 30 minutes or refrigerate for several hours.

5. To make the topping, combine the flour, sugar, baking powder, and salt in a bowl. If using lard or poultry fat, cut in the fat with two knives, a pastry blender, or your fingertips until the mixture has the consistency of coarse sand. If using tallow, combine the tallow and ½ cup of the flour mixture in a food processor and process until the mixture looks like tiny pebbles. Add the remaining flour mixture and pulse until blended. Transfer to the bowl. Add the milk and stir in with a fork. Turn out the dough onto a lightly floured surface and shape into a ball. (At this point, both the filling and dough can be refrigerated for up to 2 hours before baking.)

6. When you are ready to bake, preheat the oven to 400°F.

7. Press or roll out the dough gently into a shape that will be large enough to cover the peaches. Transfer the dough carefully to the baking dish, sealing the edges to the dish. Sprinkle with the remaining 1 tablespoon sugar. With the tip of a sharp knife, cut three or four vents in the top of the dough to allow steam to escape.

8. Bake for 35 to 40 minutes, until the biscuit topping is golden. Serve hot, if possible. It can also be served warm or at room temperature.

FRIED APPLE PIES

I did not set out to make a snack-size apple pie that tastes like the ones you get at McDonald's, but my kids tell me that my hand pies bear an uncanny resemblance. I'm not sure if that is good or bad — but I do know that these are remarkably free from grease, are just as good on the second day as on the first day (briefly reheating in the microwave is recommended), and are likely to be a real crowd-pleaser at your house, too.

Makes 8 hand pies

FILLING

¼ cup any animal fat (except bacon grease; poultry fat is recommended)

⅓ cup firmly packed light or dark brown sugar, plus more as needed

2 cups peeled, cored, and chopped apples

½ teaspoon ground cinnamon

¼ teaspoon freshly grated nutmeg

1 tablespoon unbleached all-purpose flour, mixed with 2 tablespoons water

PASTRY

2¾ cups unbleached all-purpose flour, plus more for dusting

2 tablespoons granulated sugar, plus more for sprinkling

1 teaspoon fine sea salt

⅔ cup (4.4 ounces/125g) any animal fat (except bacon grease)

1 egg

½ cup ice water, plus more as needed

About 1 cup (7 ounces/200g) lard or tallow, for frying

1. To make the filling, melt the fat in a skillet over medium heat. Add the brown sugar and stir until melted. Add the apples, cinnamon, and nutmeg, and bring to a simmer. Reduce the heat and simmer until the apples are tender and have given up their juice, about 10 minutes. Stir in the flour mixture; it should thicken immediately. Remove from the heat and chill while you make the dough.

Recipe continues on next page

Fried Apple Pies, continued

2. To make the dough, whisk together the flour, granulated sugar, and salt in a large bowl. If you are working with lard or poultry fat, cut in the fat with a pastry blender or two knives or rub in with your fingertips until the mixture resembles coarse sand. If you are working with tallow, combine the tallow and ½ cup of the flour mixture in a food processor and process until the mixture looks like tiny pebbles. Add the remaining flour mixture and pulse until well blended. Transfer to the bowl. Make a well in the center. Beat the egg with the ice water in a bowl and pour into the flour. Mix the flour into the liquids until the dough comes together and all the flour is moistened, adding additional water by the teaspoon if needed. Gather the dough into a ball and knead a little until mostly smooth. Wrap in plastic wrap and chill for at least 30 minutes, or up to overnight.

3. Line a sheet pan with parchment paper. Divide the dough in half. Roll out each half into a round with a thickness of ⅛ to ¼ inch. Using a 2½-inch round cutter, cut the dough into as many rounds as you can. You will need to gather up all the trimmings and reroll to make 16 rounds. Place on the prepared sheet pan.

4. Place about 1 tablespoon of the cooled filling in the center of eight of the rounds. Barely wet the edges of the dough with water and place a reserved round on top of each. Lightly pressing down on the edges, seal the edges with the tines of a fork.

5. Melt the fat in a large skillet and heat to 350°F. You will need at least ½ inch of fat. Fry the hand pies until browned on both sides, about 4 minutes per side. Remove from the skillet, drain on paper towels, and sprinkle with sugar while still warm. Let cool for at least 10 minutes; serve warm or completely cooled.

LARD VS. LEAF LARD

When a recipe calls for lard — in this book and others — rendered lard from any part of the pig is fine, including leaf lard. (The only lard to avoid is the shelf-stable supermarket lard.) But when a recipe — almost always a pastry or dessert recipe — specifies "leaf lard," only leaf lard will do. The crystalline structure of leaf lard is slightly different from rendered lard from elsewhere on the pig, which in turn affects the texture of the finished dish.

APPLE PIE

I have been obsessed with making a perfect apple pie for years. I want the crust flaky and tender. Well, I probably solved that with a lard crust, though a poultry fat or tallow crust works fine. Then I want a juicy filling, but it can't be runny or gummy. Cooking and thickening the apples before putting them into the crust, a make-ahead technique, solves the runniness problem because you know exactly what you are putting into the crust. This works particularly well with Granny Smith apples, which retain a pleasing firmness. Other varieties may soften more; I leave it to you to find the apple or combination of apples you like best.

Serves 8

> Double-Crust Pastry made with any animal
>   fat (except bacon grease, pages 270–72)
> 4 Granny Smith or other large apples
> 1 tablespoon lemon juice
> 2 tablespoons any animal fat (except bacon
>   grease)
> ¾ cup firmly packed light or dark brown sugar
> 2 teaspoons ground cinnamon
> ½ teaspoon freshly grated nutmeg
> ¼ cup unbleached all-purpose flour
> ¼ cup water, plus more as needed
> 1 teaspoon milk
> 1 tablespoon granulated sugar

1. Keep the pastry refrigerated while you make the filling.

2. Peel, core, and slice the apples. Toss them in a bowl with the lemon juice to prevent browning.

3. Melt the fat in a large skillet over medium heat. Add the brown sugar, cinnamon, nutmeg, and apples. Cook until the brown sugar is melted and the apples have given up their juice, about 10 minutes. Make a slurry with the flour and some water, starting with ¼ cup water and adding more as needed. Add this to the skillet; the juices should thicken immediately. Cook for 2 minutes longer. (At this point, the filling can be refrigerated for up to 1 day.)

4. Preheat the oven to 425°F with a rack in the lower third of the oven.

5. Remove the pastry from the refrigerator. Divide in half, with one piece slightly larger than the other. On a lightly floured surface, roll out the larger half to a round about ⅛ inch thick. Fit into a 9-inch pie pan, tucking the pastry into the bottom edges, and trim to leave a 1-inch overhang.

6. Spoon the filling into the pastry with a slotted spoon, leaving behind any excess liquid.

Recipe continues on page 239

Apple Pie, continued

7. Roll out the remaining pastry to a round ⅛ inch thick. Fold the dough in half over the rolling pin, lift off the work surface, and place on top of the pie. Trim to leave a 1-inch overhang and fold and tuck the overhang under the bottom crust so that it all fits inside the pie pan. Crimp the edges with a fork or make a fluted pattern with your fingers. Cut several decorative slits in the top crust to allow steam to escape.

8. Bake the pie in the lower third of the oven for 20 minutes. Reduce the heat to 350°F and continue to bake for 30 minutes longer. Remove the pie from the oven, brush with the milk, and sprinkle with the granulated sugar. Return to the oven and bake for 10 to 15 minutes longer, until the crust is golden and the juices are bubbly.

9. Cool the pie on a wire rack and serve warm or at room temperature.

DUCK FAT–CARAMELIZED APPLE TART

This tart is an easy one to whip together but looks quite impressive. The pastry is a standard pie pastry, rolled out and cut into two equal rectangles. The filling is the Duck Fat–Caramelized Apples on page 252. Throw the two together to make a lovely tart you can prepare ahead, then bake while dinner is served.

Serves 6

> Double-Crust Pastry made with any animal fat (except bacon grease, pages 270–72)
> Duck Fat–Caramelized Apples (page 252; see step 1)
> 1 tablespoon light or dark brown sugar
> 1 tablespoon unbleached all-purpose flour
> 1 egg beaten with 1 tablespoon water, for egg wash
> 2 tablespoons granulated sugar

1. Divide the pastry into two equal halves (a scale is helpful here), form into flattened rectangles, wrap each in plastic wrap, and refrigerate for at least 30 minutes, or up to 1 day.

2. Prepare the caramelized apples according to the recipe directions. Before taking them off the heat, stir in the 1 tablespoon brown

Recipe continues on next page

Duck Fat–Caramelized Apple Tart, continued

sugar and the 1 tablespoon flour. Transfer to another container and chill.

3. Line a half sheet pan with parchment paper. On a well-floured surface, roll out each pastry rectangle into a rectangle that is about ⅛ inch thick. The exact dimensions of the rectangle are not important, but I generally roll out each to a rectangle that is 6 to 8 inches wide and 16 inches long. Use a bench knife or pizza cutter and ruler to even out the edges, and transfer one of the rectangles to the prepared sheet pan. Top the pastry with the chilled apple filling, leaving a 1-inch border all around. Top with the second rectangle of pastry. Fold or roll the edges of the pastry sheets up to form a raised edge. Crimp the edges.

4. If you are going to delay baking, cover lightly with plastic wrap and return to the refrigerator for up to 2 hours.

5. To bake, preheat the oven to 425°F. Brush the pastry with the egg wash, sprinkle with the granulated sugar, and cut three or four steam vents in the top.

6. Bake for 25 to 30 minutes, until the crust is golden brown. Let cool for at least 20 minutes before serving so the filling isn't burning hot.

EASY FRENCH APPLE TART

In this tart, a sweet pastry crust and a simple cream cheese filling lie under an apple filling that's used in another tart on page 239. The crust — made with the animal fat of your choosing — is patted into place, the filling is whipped together in a food processor, and this impressive dessert couldn't be simpler to put together.

Serves 8

    Duck Fat–Caramelized Apples (page 252)
 4 ounces cream cheese
 ½ cup confectioners' sugar
 1 teaspoon pure vanilla extract
    Sweet Tart Pastry made with any animal fat
       except bacon grease (pages 274–76), fully
       baked and cooled

1. Make sure the apples are cooled; refrigerate for at least 1 hour if you've just made them.

2. Combine the cream cheese, confectioners' sugar, and vanilla in a food processor and process until well blended and smooth.

3. Spread the cream cheese filling in the tart crust. Top with the apples. The tart can be served immediately or held in the refrigerator for up to 6 hours. The tart is best on the day it is made, but leftovers will still be enjoyable on the second and third days.

BLUEBERRY GALETTE

Blueberry season begins in Florida and moves up the East Coast and into Canada. My bushes begin yielding in mid-July and continue through August, producing quite an abundance. This galette — or crostata if you are feeling Italian — is a rustic tart that looks beautiful no matter how it is shaped, or even what fruit you use. Throughout the fall and winter, I swap out the blueberries in favor of apples or pears, and use cinnamon and brown sugar instead of lemon and white sugar. The lard pastry handles like a dream, rolling out quickly and easily. A topping of vanilla ice cream does not go amiss.

Serves 6–8

Double-Crust Pastry Made with Lard
  (page 270)
4 cups fresh blueberries
½ cup sugar, plus 1 tablespoon for sprinkling
2 tablespoons cornstarch
1 teaspoon very finely grated lemon zest
2 tablespoons lemon juice
1 egg beaten with 1 tablespoon water, for
   egg wash

1. Chill the pastry for at least 30 minutes.

2. Preheat the oven to 350°F with a rack in the center of the oven.

3. Roll out the pastry on a sheet of parchment paper to a thickness of about ⅛ inch and at least 12 inches in diameter. You can trim it to make a perfectly round 12-inch circle of dough, but it isn't necessary for the rustic appearance of this tart. Transfer the pastry, still on the parchment paper, to a half sheet pan. It is okay if you have to drape the edges of the pastry over the rim of the pan.

4. Combine the berries, ½ cup sugar, cornstarch, lemon zest, and lemon juice in a bowl and toss gently to mix with a silicone spatula. Dump the berry mixture onto the center of the pastry, leaving about 2 inches around the edge. Spread evenly. Fold the pastry edge over the filling, pleating as you go. Brush the folded-over edge with the egg wash and sprinkle with the remaining 1 tablespoon sugar.

5. Bake for 55 to 60 minutes, until the crust is golden and the center is bubbly. Place the sheet pan on a rack and let cool. When cool enough to handle, use two metal spatulas to transfer the galette to a serving plate or cutting board. Slice it and serve it warm or at room temperature.

FRESH FRUIT TART

The fruit is up to you, but berries, peaches, and citrus are particularly lovely here. The filling is a classic pastry cream. In bakeries, these tarts are painted with melted jelly so they sparkle and hold up for hours under bright lights. At home you can forgo the melted jelly finish if you prefer a more rustic look.

Serves 6

- 4 egg yolks, beaten
- ⅛ teaspoon fine sea salt
- ½ cup sugar
- ¼ cup cornstarch
- 2 cups milk
- 2 tablespoons any animal fat (except bacon grease) or butter
- 1 teaspoon pure vanilla extract
- Sweet Tart Pastry made with any animal fat (except bacon grease, pages 274–76), fully baked and cooled
- 2 cups fresh whole berries or sliced fruit
- ⅓ cup apple or currant jelly (optional)
- 1 tablespoon water (optional)

1. Beat together the egg yolks and salt in a bowl. Stir together ¼ cup of the sugar and the cornstarch in another bowl. Stir the sugar mixture into the egg yolks and mix until smooth.

2. Stir together the milk and the remaining ¼ cup sugar in a small saucepan over medium heat and bring to a boil. Slowly drizzle the milk mixture into the egg mixture, stirring constantly. When all the milk is added, return the whole mixture to the saucepan. (This step tempers the eggs and prevents them from curdling.)

3. Bring the milk and egg mixture to a boil, stirring constantly, over medium heat; the mixture will thicken. As soon as it comes to a boil, remove from the heat and stir in the fat and vanilla.

4. Transfer the mixture to a bowl, cover with plastic wrap directly on the surface to prevent a skin from forming, and chill in the refrigerator for at least 2 hours, or up to 1 day.

5. To assemble the tart, spoon the chilled pastry cream into the tart crust and smooth the top. Arrange the fruit on top. To glaze, if desired, combine the jelly with the water in a small saucepan and heat until just melted. With a pastry brush, paint the fruit with the jelly. Let set for about 10 minutes. Serve or refrigerate for a few hours; the tart is best enjoyed on the day it is made.

CHOCOLATE-PEAR GALETTE

Chocolate with any fruit makes a wonderful pairing, and this quickly assembled, rustic tart is delicious proof.

Serves 6–8

Double-Crust Pastry Made with Lard
   (page 270)
½ cup sugar, plus 1 tablespoon for sprinkling
2 tablespoons cornstarch
1 teaspoon fine sea salt
1 teaspoon ground cinnamon
4 cups peeled and sliced large ripe pears
   (6–8 pears)
4 ounces semisweet chocolate, grated or finely
   chopped
1 egg beaten with 1 tablespoon water, for
   egg wash

1. Chill the pastry for at least 30 minutes.

2. Preheat the oven to 350°F with a rack in the center of the oven.

3. Roll out the pastry on a sheet of parchment paper to a thickness of about ⅛ inch and at least 12 inches in diameter. You can trim it to make a perfectly round 12-inch circle of dough, but it isn't necessary for the rustic appearance of this tart. Transfer the pastry, still on the parchment paper, onto a half sheet pan. It is okay if you have to drape the edges of the pastry over the rim of the pan.

4. Mix the ½ cup sugar, cornstarch, salt, and cinnamon in a bowl. Add the pears and toss to coat.

5. Sprinkle about two-thirds of the chocolate on the dough, leaving about 2 inches around the edge. Dump the pears over the chocolate and spread out in an even layer. Fold the dough edge over the filling, overlapping and pleating the edges. Brush the folded-over edge with the egg wash and sprinkle with the remaining grated chocolate and the remaining 1 tablespoon sugar.

6. Bake for about 55 minutes, until the crust is golden and the center is bubbly. Place the sheet pan on a rack and let cool for at least 15 minutes. Use two metal spatulas to transfer the galette to a serving plate or cutting board. Slice it and serve it warm or at room temperature.

PEACH TART

Here's a tart you can whip up for a summer dessert in no time — you can even substitute canned peaches for the fresh ones if peeling seems like too much work. To peel peaches, slip into boiling water and blanch for 30 to 60 seconds. The skins should then slip right off.

Serves 6

> Sweet Tart Pastry made with any animal fat
> (except bacon grease, pages 274–76)
> 4 cups peeled, pitted, and sliced peaches
> ½ cup sugar
> ¼ cup cornstarch
> ¼ cup gingersnap crumbs

1. Press the pastry into a 9-inch tart pan with a removable bottom. Place in the freezer for about 30 minutes.

2. Preheat the oven to 350°F with a rack in the middle of the oven.

3. Combine the peaches, sugar, and cornstarch in a bowl and stir to blend. Spoon the peaches into the tart shell. Sprinkle the gingersnap crumbs evenly on top.

4. Place the tart pan on a sheet pan and bake for 50 to 55 minutes, until the crust is golden and the filling is bubbling. Let cool for 10 minutes before removing the sides of the tart pan. Serve warm or at room temperature.

JAM TART

Sometime in the spring each year, I become aware that I have made more jam than needed. There are stacks on the shelf, but I know I will end up making jam again. That's when it's time to whip out recipes that use up jam, like this very easy tart adapted from David Lebovitz. It has a cookie dough crust that is pressed into shape, so no rolling is required. Vanilla ice cream is a lovely accompaniment.

Serves 8–12

> ½ cup (3.5 ounces/100g) any animal fat (except bacon grease)
> ½ cup sugar, plus more for sprinkling
> ½ teaspoon pure almond or vanilla extract
> 2 eggs, 1 whole and 1 separated
> 1½ cups unbleached all-purpose flour
> ½ cup stone-ground cornmeal or polenta
> 2 teaspoons baking powder
> ½ teaspoon fine sea salt
> 1¾ cups blueberry, apricot, raspberry, or other jam
> 1 tablespoon water

1. Blend the fat and sugar in a food processor until well combined. Add the vanilla, the whole egg, and the egg yolk, and process until just blended. Reserve the egg white.

Recipe continues on next page

2. Whisk together the flour, cornmeal, baking powder, and salt in a bowl. Add to the food processor and pulse just until the mixture comes together.

3. Divide the dough into two pieces, one with about two-thirds of the dough and the other with about one-third of the dough. Shape the larger piece into a disk and wrap in plastic. Roll the smaller piece into a log about 2 inches in diameter and wrap in plastic. Chill both in the refrigerator for at least 30 minutes, or up to 1 day.

4. Preheat the oven to 375°F. Very lightly grease a 9- or 10-inch tart pan with a removable bottom, a springform pan, or a regular pie dish.

5. Remove the dough from the refrigerator. With the heel of your hand, press the dough disk into the bottom and up the sides of the prepared pan, patting it out evenly. Spread the jam evenly over the dough.

6. Slice the log of dough into ¼-inch-thick disks and lay them over the jam in a decorative pattern. Lightly beat the reserved egg white with the water and brush onto the top crust. Sprinkle generously with additional sugar, using at least 2 tablespoons.

7. Bake for 20 to 25 minutes, until the pastry is golden brown. Let cool on a wire rack and serve at room temperature.

CARROT CAKE CUPCAKES

It's rare to find a carrot cake recipe that doesn't use vegetable oil instead of butter in its formula. Why? Could be that its burgeoning popularity in the 1970s made people think that vegetable oil (and sometimes whole wheat) was a healthful complement to the carrots? Lard adds moisture to the cupcakes without adding any porky notes. These are absolutely delicious.

Makes 24 cupcakes

CUPCAKES

    1 pound carrots, peeled
 2½ cups unbleached all-purpose flour
 1¼ teaspoons baking powder
    1 teaspoon baking soda
    1 teaspoon fine sea salt
    1 teaspoon ground cinnamon
  ½ teaspoon ground ginger
  ¼ teaspoon freshly grated nutmeg
    1 cup (7 ounces/200g) leaf lard
    1 cup granulated sugar
  ½ cup firmly packed light or dark brown sugar
    4 eggs
    2 teaspoons pure vanilla extract

Recipe continues on page 248

Carrot Cake Cupcakes, continued

   1 cup crushed pineapple, well drained

   1 cup walnuts, chopped

FROSTING

   4 ounces cream cheese, softened

   ½ cup (1 stick) butter, softened

   2 cups confectioners' sugar

   1 teaspoon pure vanilla extract

1. Preheat the oven to 350°F with a rack in the center of the oven. Line two 12-cup muffin pans with cupcake liners.

2. Grate the carrots in a food processor fitted with the grating disc. Remove the carrots from the bowl, but do not wash the bowl. Set the carrots aside.

3. Whisk together the flour, baking powder, baking soda, salt, cinnamon, ginger, and nutmeg in a large bowl.

4. Combine the lard with the granulated and brown sugars in the food processor bowl and process until well blended. Add the eggs, one at a time, processing to blend. Add the vanilla and the flour mixture and process until well blended.

5. Transfer the batter to a clean bowl. Fold in the carrots, pineapple, and walnuts.

6. Scoop the batter into the cupcake liners, filling each one just barely full. Bake for 20 to

25 minutes, until the tops are golden brown and a toothpick inserted into one of the cakes comes out clean. Let cool for 10 minutes, then turn out onto wire racks and cool completely.

7. To make the frosting, combine the cream cheese and butter in the bowl of an electric stand mixer fitted with the whip attachment and beat until thoroughly combined. Add the confectioners' sugar and vanilla and beat until smooth. Spread the frosting over the top of each cooled cupcake.

POUND CAKE

Similar to the Blueberry Muffins on page 231, this is a proof-of-concept recipe. Would we taste the lard in such a classic cake made of so few ingredients? Would we miss the butter? No and no. Although lard makes a slightly denser cake (pound cake is already dense) with a tighter crumb, this tastes and feels like a pound cake made with butter. If it lasts very long, you may notice that the cake retains its moist texture slightly longer than one made with butter.

Serves 12

   1 cup (7 ounces/200g) leaf lard

1¾ cups granulated sugar

   5 eggs

1½ teaspoons pure vanilla extract, or 1 tablespoon finely grated lemon zest plus
1 tablespoon lemon juice
1 teaspoon fine sea salt
2 cups unbleached all-purpose flour
Sifted confectioners' sugar, for dusting

1. Preheat the oven to 350°F. Lightly grease a
5- by 9-inch loaf pan and line it with parchment paper.

2. Beat the lard in the bowl of an electric mixer
fitted with the whip attachment until very
light and creamy (this is important, so be
patient). Add the granulated sugar gradually
and continue beating for about 5 minutes
longer, until the mixture is very fluffy. Beat in
the eggs, one at a time, beating well after each
addition. Keep beating until the color light-
ens. Add the vanilla and salt and beat them in.

3. Fold in the flour by hand, mixing just until
the batter is smooth and blended (don't try
to save time by beating it in, lest your cake
overrise and then fall). Scrape the batter
into the prepared pan and smooth the top.

4. Bake for about 1½ hours, until a cake tester
inserted in the center of the cake comes
out clean and an instant-read thermometer
registers at least 210°F. If the cake starts
browning too much on top, cover loosely
with aluminum foil.

5. Cool on a wire rack for about 10 minutes.
Remove the cake from the pan, peel off the
parchment, and cool completely. Dust with
confectioners' sugar right before serving.

APPLE STREUSEL COFFEE CAKE

Here's a classic coffee cake, with a filling of apples
adding a little magic. You won't taste the lard or
poultry fat, and you won't miss the butter.

Serves 12

STREUSEL TOPPING
½ cup unbleached all-purpose flour
½ cup granulated sugar
1 teaspoon ground cinnamon
½ teaspoon fine sea salt
1 cup thinly sliced almonds
¼ cup (1.7 ounces/50g) lard or poultry fat,
melted
2 tablespoons pure maple syrup

FILLING
2 large apples, peeled, cored, and finely
chopped
1 cup firmly packed light or dark brown sugar
2 teaspoons ground cinnamon

Recipe continues on next page

Apple Streusel Coffee Cake, continued

CAKE

    3 cups unbleached all-purpose flour
    2½ teaspoons baking powder
    1½ teaspoons fine sea salt
    ½ teaspoon freshly grated nutmeg
    9 tablespoons (3.8 ounces/115g) leaf lard,
        at room temperature
    1 cup granulated sugar
    2 tablespoons light or dark brown sugar
    2 eggs
    1 teaspoon pure vanilla extract
    1¼ cups buttermilk

1. Preheat the oven to 350°F. Lightly grease a 9- by 13-inch baking dish.

2. To make the topping, whisk together the flour, sugar, cinnamon, and salt in a bowl. Stir in the almonds. Add the melted fat and maple syrup and stir until well combined. Set the topping aside.

3. To make the filling, mix together the apples, brown sugar, and cinnamon in a bowl. Set aside.

4. To make the cake, whisk together the flour, baking powder, salt, and nutmeg in a bowl.

5. Beat together the lard and granulated and brown sugars in the bowl of a stand mixer fitted with the paddle attachment on medium speed until well combined and smooth. Add the eggs, one at a time, beating well after each addition. Beat in the vanilla. Add the flour mixture in thirds, alternating with the buttermilk, beating gently to combine after each addition.

6. Pour slightly more than half the batter into the prepared pan, spreading it all the way to the edges. Sprinkle the filling evenly on the batter. Spread the remaining batter on top of the filling. Use a table knife to gently swirl the filling into the batter, as though you were making a marble cake. Sprinkle the topping over the batter in the pan.

7. Bake the cake for 50 to 55 minutes, until the top is golden and a toothpick inserted into the center comes out clean. Cool for at least 20 minutes on a wire rack before cutting and serving.

ZUCCHINI SPICE CAKE

Zucchini hit the American palate at about the same time Americans started fearing animal fats. So a lot of zucchini cakes and quick breads call for using vegetable oil in the batter. Coincidence? I don't think so. Lard makes it better. A classic cream cheese frosting makes it even better.

Serves 12

CAKE

- 1 pound zucchini
- 1 teaspoon fine sea salt
- 2½ cups unbleached all-purpose flour
- 1¼ teaspoons baking powder
- 1 teaspoon baking soda
- 1 teaspoon ground cinnamon
- ½ teaspoon ground ginger
- ¼ teaspoon freshly grated nutmeg
- ¼ teaspoon ground allspice
- 1 cup (7 ounces/200g) leaf lard, at room temperature
- 1 cup granulated sugar
- ½ cup firmly packed light or dark brown sugar
- 4 eggs
- 2 teaspoons pure vanilla extract
- 1 cup walnuts, chopped

FROSTING

- 8 ounces cream cheese, softened
- ½ cup (1 stick) butter, softened
- 2½ cups confectioners' sugar
- 1½ teaspoons pure vanilla extract

1. To make the cake, grate the zucchini in a food processor fitted with the grating disc. Remove the zucchini from the bowl, but do not wash the bowl. Combine the zucchini and salt in a colander and let drain for at least 30 minutes.

2. Preheat the oven to 350°F with a rack in the center of the oven. Grease two 8- or 9-inch round pans. Line the bottoms with parchment paper.

3. Whisk together the flour, baking powder, baking soda, cinnamon, ginger, nutmeg, and allspice in a large bowl.

4. Combine the lard with the granulated and brown sugars in the food processor bowl and process until well blended. Add the eggs, one at a time, processing to blend. Add the vanilla and the flour mixture and process until well blended.

5. Transfer the batter to a clean bowl. Fold in the zucchini and walnuts.

6. Divide the batter between the prepared pans, smooth the tops, and bake for 35 to 45 minutes, until the tops are golden brown and a toothpick inserted comes out clean. Let cool for 10 minutes, then turn out onto wire racks and cool completely.

Recipe continues on next page

7. To make the frosting, combine the cream cheese and butter in the bowl of a stand mixer fitted with the whip attachment and beat until thoroughly combined. Add the confectioners' sugar and vanilla and beat until smooth.

8. Level the tops of the cakes as needed. Choose one to be the bottom layer and place on a plate. Spread about one quarter of the frosting on top. Place the second layer on top of the frosted layer. Spread the remaining frosting over the top and sides of the cake, covering it completely.

DUCK FAT–CARAMELIZED APPLES

The first time I made these delicious apples, I folded them into rice pudding. And the pudding was so delicious, I had to think of more and more uses for the apples. Serve them folded into winter squash (page 196), in a pastry shell (pages 274–76), on top of ice cream, or on top of pancakes and waffles. Or just stand by the stove and enjoy them by the spoonful.

Serves 4–6

- ¼ cup (1.7 ounces/50 grams) duck fat (or any poultry fat)
- ¼ cup firmly packed light or dark brown sugar, or as needed
- ¼ teaspoon ground cinnamon
- 6 cups peeled, cored, and diced apples
- 1 tablespoon unbleached all-purpose flour, mixed with 2 tablespoons water (if using in a tart)

1. Melt the duck fat in a large skillet over medium heat. Add the brown sugar and cinnamon and stir until the brown sugar melts into the fat, about 2 minutes. Add the apples, stir to coat in the sugar mixture, and continue to cook, stirring frequently, until the apples are tender, about 10 minutes, depending on the variety. If the apples are to be used as a tart filling or topping, stir in the flour mixture and continue to cook until the filling is thick, about 5 minutes longer.

2. Taste and add more brown sugar, if needed. Stir until the sugar dissolves, then remove from the heat. Serve warm or cooled. Chill before using in a pastry.

THE ROLE OF FATS IN BAKED GOODS

Fats play an important role in providing flavor and a tender texture to baked goods. They also aid in the browning process of many items, giving them an appealing golden brown color.

Fat has the unique ability to absorb and preserve flavors. Many flavors are soluble in fat, but not in water. Fat coats the tongue and allows flavors to linger, which also enhances how we enjoy foods. Saturated fats, like tallow and lard, are solid at room temperature, which makes them perfect for using in solid foods like chocolate and frosting. By contrast, vegetable oils are liquid at room temperature, which makes them suitable for use in products like salad dressings.

Normally, as bread dough is kneaded the gluten (wheat protein) begins to join and form long elastic strands, which give strength and a chewy texture to the bread. When fat is added to dough, as in biscuits and piecrusts, the fat provides a physical barrier that prevents the long gluten strands from forming, thus keeping the dough tender. As the fat melts, it forms pockets that become those flaky layers.

CHAPTER 8

BASICS

SIMPLE ROAST DUCK

Sometimes you just want a simple roasted duck — no complicated steps, no hard-to-find ingredients. That's what this recipe is about. There is one trick that, while not completely standard, is absolutely necessary — to spatchcock or butterfly the bird first. This step of removing the backbone and flattening the breast maximizes the amount of fat you can trim off the bird for rendering and guarantees that the breast and legs all cook at the same time. After cooking in the oven at a fairly low temperature, run the bird under the broiler to crisp the skin. Pretty simple, no?

Serves 4

> 1 (5- to 7-pound) duck
> Salt and freshly ground black pepper
> Hoisin sauce, barbecue sauce, or orange marmalade mixed with equal parts soy sauce, for glazing

1. Preheat the oven to 325°F. Fill a kettle with water and bring to a boil.

2. Reach into the cavity of the bird and remove the neck, heart, gizzard, liver, and any loose fat. Put the liver in an airtight container and refrigerate or freeze to make a pâté. Set aside the neck, heart, and gizzard to make stock. Set aside the fat. Trim the excess skin from the neck area and add to the loose fat.

3. To spatchcock the bird, place it breast side down on a cutting board. Use poultry shears or a cleaver to cut along both sides of the backbone, beginning at the tail end. If you hit a tough spot, try cutting with just the tip of the shears. A rubber mallet tapped on the blade of the cleaver may also be helpful. Trim the skin off the backbone and add to the fat pile. The end of the backbone, sometimes irreverently called the pope's nose or the parson's nose, can be removed and added to the container that holds the duck skin and fat. Add what remains of the backbone to the stock pile.

4. Turn the bird breast side up. Trim off the wing tips and add to the stock pile. Place both hands on the breast bone and push down hard until you feel the breast give way and the carcass flatten. There will be extra skin and fat hanging off the thighs. Trim this and any extra fat you can. Refrigerate or freeze the skin and fat to render at another time. Refrigerate or freeze the bones and organs for stock. To make the broth, follow the recipe on page 153.

5. Put the bird in a roasting pan. Pour the boiling water over the bird to tighten the skin. Dump out the water and pat the bird dry. Dry the roasting pan. Rub the bird all over with salt and pepper and return to the roasting pan, skin side up. Pull the thighs

outward so the bird lies flat, with the wings facing inward.

6. Roast for about 1½ hours, until the bird registers 165°F in the thickest part of the thigh. Remove the bird from the oven and pour off and reserve any accumulated fat and juices. (The reserved fat and juices can be stored in the refrigerator. The fat will rise to the top of the container and can be spooned off and used to cook with. The juices can be added to stocks or sauces to enrich the flavor; use or freeze within 1 week.)

7. Turn on the oven broiler. Paint the bird with the glaze. Place under the broiler and broil until the skin is crispy and colored, 3 to 5 minutes.

8. Cut the bird into quarters to serve.

MULTISTEP ROASTED GOOSE

There's no question that goose is an expensive choice for a holiday meal. Whereas bargain supermarket turkeys are often sold at a loss around holiday times, you won't find similarly priced geese. With meat that expensive, you want to cook it right. I was happy enough with a long, slow roast for a spatchcocked goose, cooked like the duck on page 256. But that method doesn't fully render out all the fat. The recipe below does render more of the fat, but it is a two-day process. The method is cobbled together from Julia Child's *The Way to Cook* and an article about roasting goose by Niki Achitoff-Gray. The end result is a lovely dark meat goose that's like a slightly gamier, tougher version of duck with lots of lovely goose fat — pure liquid gold. The carcass will yield a delicious broth.

Serves 8–10

- 1 whole goose (10–12 pounds)
- 1 lemon
- ¼ cup kosher salt
- 1 tablespoon baking powder
- 2 large onions
- 4 celery stalks
  Several sprigs of parsley, sage, and/or thyme
- ½ cup port wine
- 2 tablespoons cornstarch
  Freshly ground black pepper

1. First, prepare the goose for cooking. Inside the cavity of the bird should be the neck, gizzard, heart, and liver. Refrigerate or freeze the liver to make pâté at another time. Put the neck, gizzard, and heart in a stockpot. Trim off the wing tips and add them to the stockpot. With kitchen shears or a sharp paring knife, trim the excess skin from the goose's neck; be aggressive and trim as close to the carcass as possible. Pull out lumps

Recipe continues on next page

of pale fat from the cavity, pulling out from each end. Put the fat and skin in an airtight container and refrigerate until you are ready to render it. (You should have about 1 pound of fat and skin, which will yield about 2 cups of rendered fat.) With the bird breast side up, use your fingers to locate the wishbone at the top of the carcass on the neck end. Insert a boning knife to separate the wishbone from the rest of the carcass. Pull the bone down and away from the carcass. If you have freed it from the meat at the top of the carcass, you should be able to snap it off. (This will make the roasted bird easier to carve.) Add the wishbone to the pot that holds the neck, gizzards, and heart.

2. To tighten the skin and allow the bird to keep its shape despite the lengthy cooking time, the next step is to blanch the bird. Add water to a large stockpot to fill two-thirds and bring to a boil over high heat.

3. Grasp the legs of the goose and lower it into the water, neck end first, submerging it halfway. Keep it submerged for 1 minute. Lift the goose, allowing excess water to drain back into the pot, then transfer the bird to a work surface. Grasp the wings and submerge the other half of the goose, tail end first, in the boiling water and keep it submerged for another 1 minute. Lift the goose, allowing excess water to drain back into the pot, then transfer the goose to a work surface. Pat dry with paper towels inside and out. Reserve the blanching water.

4. Dry-brine the bird for 12 to 24 hours. This helps the skin become crispy later. Cut the lemon in half and squeeze all over the bird, inside and out. Put the lemon halves inside the carcass. In a small bowl, mix the kosher salt and baking powder together. Generously and evenly sprinkle the salt mixture all over the goose skin. Transfer the goose to a rack set in a stovetop-safe roasting pan and refrigerate, uncovered, for 12 to 24 hours.

5. While the bird dry-brines, make the stock. Coarsely chop 1 onion and 2 celery stalks and add them to the pot that holds the neck and wishbone. Add about 6 cups water and bring to a boil. Reduce the heat and simmer for 2 hours. Strain and refrigerate the stock; dispose of all the solids.

6. To steam-roast the bird, brush the salt mixture off the bird. Using the tip of a paring knife, prick the skin at ½-inch intervals all over, front and back, being sure to pierce the skin but not poke holes in the meat. Pay special attention to particularly fatty areas, beneath the wings and around the

thighs. Place the bird on a rack, breast side up, in the roasting pan. Add the herbs to the cavity. Pour 1 or 2 inches of water in the pan and bring to a boil on top of the stove (probably over two burners). Cover the pan tightly with a lid or with aluminum foil, reduce the heat, and steam for 1 hour. Remove from the heat and let cool for 20 minutes.

7. Preheat the oven to 325°F.

8. Transfer the bird to a plate or baking sheet. Remove the rack from the roasting pan. Pour off the cooking liquid — a combination of broth and fat — into one or two widemouthed quart canning jars.

9. Chop the remaining 1 onion and 2 celery stalks and add to the roasting pan. Place the goose on top of the vegetables. Pour 1 cup of the reserved blanching water over the bird. Cover tightly. Transfer the goose to the oven and roast for 1½ hours, until the meat all over the bird measures 165°F. Remove from the oven and carefully transfer the goose to a baking sheet. Spoon or pour off all the rendered fat and juices in the roasting pan into a heatproof bowl or container, being careful to leave behind any solids. Reserve the rendered fat.

10. Turn on the oven broiler. Broil the goose until the skin is browned, 5 to 10 minutes.

Remove from the oven and let stand, uncovered, for 30 minutes. Meanwhile, transfer any newly accumulated rendered fat from the roasting pan to the reserved fat. Do not wash the roasting pan.

11. In a medium saucepan, bring 2 cups of the goose stock to a simmer. Heat the roasting pan on two burners over low heat. Add 1 cup of the heated goose stock and, using a wooden spoon, scrape up any browned bits on the bottom. Bring to a boil and let boil for 30 seconds. Strain the contents of the roasting pan into the saucepan with the goose stock, scraping all the browned bits from the bottom of the pan. Continue simmering for 5 minutes.

12. Combine the port and cornstarch in a small bowl and stir until smooth. Pour into the stock and simmer for at least 5 minutes; as it simmers, the gravy will thicken. Season with salt and pepper to taste. Keep warm.

13. To carve the goose, remove the two legs, then the two wings. Cutting down from one side of the breastbone, remove the whole breast half in one piece. Repeat on the other side. Slice the breast meat on the diagonal to make medallions about ½ inch thick. Slice the meat off the thigh bones about ½ inch thick. Arrange the drumstick, wings, and sliced meat on a platter and serve, passing the gravy on the side.

CONFIT OF GIZZARDS

How many times have you discarded chicken gizzards because they are odd little organs and you have no good recipes for them? Well, waste them no more, because this is an absolutely delicious way to create flavorful, tender gizzards that make a great addition to salads, grains, stuffings, gravies — you name it. The famed Jewish cookbook author Joan Nathan adds them to Kasha Varnishkas (page 211).

Makes 2 cups gizzards plus 2 cups fat

- 1 pound gizzards or gizzards and hearts from any type of poultry
- 2 teaspoons fine sea salt
- 1 teaspoon freshly ground black pepper
- 3 garlic cloves, minced
- 3 thyme sprigs
- 3 bay leaves
- 1¼ cups (8.7 ounces/250g) poultry fat, melted
  French bread, for serving
  Whole-grain mustard, for serving
  Pickles, for serving
- 1 cup gribenes (poultry cracklings), for serving (optional)

1. Preheat the oven to 275°F.

2. Clean the gizzards, if needed. (Store-bought gizzards have been cleaned. Gizzards from freshly slaughtered poultry should be chilled before cleaning. To clean them, trim away all the fat, slice open until you can see white, then pull apart the lobes to expose the sac containing the grit and stones. Discard the sac.) Chop the gizzards and hearts into ½-inch pieces.

3. Put the gizzards in a small baking dish and add the salt, pepper, and garlic. Toss to mix. Tuck the thyme sprigs and bay leaves among the gizzards. Pour the melted fat over all. Cover and bake for 2 to 3 hours, until tender but still slightly chewy.

4. Discard the thyme and bay leaves. Drain the gizzards, reserving the fat. Use at once, or store in glass canning jars, with enough fat to completely submerge the gizzards. If necessary, add more melted poultry fat, melted lard, or olive oil to cover. Refrigerate until you are ready to use. The gizzards will keep for up to 2 weeks covered in the fat in the refrigerator or 6 months in the freezer. The reserved fat will keep for 2 to 3 months in the refrigerator and up to 1 year in the freezer.

5. Serve with French bread, mustard, pickles, and gribenes (if desired).

DUCK THREE WAYS: PAN-SEARED BREASTS, DUCK CONFIT, DUCK BROTH

Although supermarkets do sell duck breasts and duck legs separately, if you need four duck legs for confit, buying two whole birds makes more sense than just buying legs; otherwise, you lose out on the opportunity to render all that luscious duck fat. So the indulgence of two ducks can turn into several different meals. The four confit duck legs can make a few different meals, such as roasted duck with potatoes (page 162) or salad with duck confit (page 129); the breasts can make one glorious meal of pan-seared duck breasts; and the bones make terrific stock. The confit wings are either the cook's reward for the work or they offer just enough meat to garnish a French *garbure* (page 161). Add in the pâté you can make from the two livers and the duck fat you will render, and you can call buying a couple of ducks a good investment.

Makes 2–4 meals for 4, plus broth and duck fat

2 (6- to 7-pound) ducks

CONFIT

¼ cup kosher salt

4 garlic cloves, smashed

4 thyme sprigs

6 bay leaves

Coarsely ground black pepper

About 3 cups poultry fat

BROTH

4 celery stalks, chopped

2 onions, chopped

1 bunch parsley

About 10 cups water

BREAK DOWN THE DUCKS

1. Work with one duck at a time. Inside the body cavity, you should find the neck, giblets, heart, and liver. Refrigerate or freeze the liver for pâté. The heart and giblets can be used for the stock along with the neck, or they can be made into confit with the legs. If you are lucky, there will be a thick pad of fat to remove from the cavity as well; save this for rendering. There should also be a length of skin that once covered the neck. Cut this off and reserve for rendering.

Recipe continues on next page

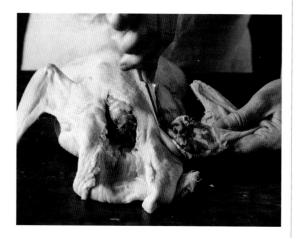

2. Place the duck breast side down on a cutting board. Using a small knife, make a cut underneath one of the wings, grab the wing with your free hand, and pull it away from the body. Cut under and around the joint to remove the wing. Repeat to remove the second wing. Chop off the wing tips to use for stock.

3. Turn the duck breast side up. Make a cut between one leg and the body. Grab the duck leg and pull it away from the body to expose the joint. With the knife angled flush against the carcass, cut under and around the joint. Pull the leg away from the body and cut down the back of the duck to detach the leg. Repeat with the remaining leg. Trim away and reserve the excess fat from each leg and set the legs aside for confit.

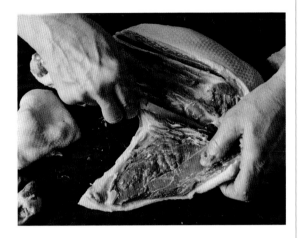

4. Next you will need to peel the breast meat from the bones to allow the breasts to sit flat in a skillet and cook evenly. With your fingers, feel for the thin breastbone that runs down the length of the breast from the neck cavity to the tail. Working slowly and using the breastbone as your guide, cut down the length of bone about 1 inch deep until you reach the rib cage. Gently peel the breast away from the carcass, sliding your knife along the rib cage toward the tail. You want to see as little red flesh remaining on the bone as possible. Work back up toward the neck end and cut around the wishbone. Gently free the breast by sliding the knife against the carcass. Repeat with the other breast.

5. Cut off and reserve the tail and neck fat and any fat from the carcass. Peel off the skin from the backbone. Turn each breast skin side down and trim away and reserve the excess fat and skin for a more elegant presentation.

6. You now have two bone-in legs and two wings for confit, two boneless breasts to pan-sear, and a partial carcass and two wing tips for stock. You should also have a pile of skin and fat to render. Repeat to break down the second duck.

Recipe continues on next page

SALT THE LEGS AND WINGS FOR CONFIT

7. Sprinkle 1 tablespoon of the salt in the bottom of a dish or plastic container large enough to hold the duck pieces in a single layer. Arrange the duck, skin side up, over the salt mixture. Sprinkle the duck with the garlic, thyme, bay leaves, and the remaining 3 tablespoons salt, as well as a few grinds of pepper. Cover and refrigerate for 1 to 2 days.

RENDER THE FAT

8. Chop the fat and skin into small pieces, about ½ inch in size. Place in a large heavy saucepan. Turn the heat to medium and slowly render the fat, stirring frequently. The fat should render out and the cracklings (or gribenes) turn brown in 1 to 1½ hours. Strain the fat, refrigerate, and reserve the cracklings (see page 53). You should have 2 to 3 cups of rendered fat, plus about 2 cups of cracklings.

MAKE THE BROTH

9. To make the broth, chop the carcass into manageable pieces and put in a large heavy saucepan with the celery, onions, parsley, and water. Bring just to a boil, then reduce the heat and simmer gently for 3 to 4 hours. Strain out and discard the solids and refrigerate or freeze the broth. You should have about 4 quarts.

PAN-SEAR THE BREASTS

10. Two tricks guarantee success when cooking duck breasts: scoring the skin and the thick layer of fat covering the top of each breast to allow the fat to render out, and starting to cook the breasts in a cold pan to allow the fat to render out before the skin begins to brown.

11. To score the skin, arrange the first breast skin side up. Using a sharp knife, make diagonal cuts about ¼ inch apart through the skin and fat without piercing the flesh. Turn the breast 45 degrees and cut crosswise incisions ¼ inch apart to make a diamond pattern. Repeat with the remaining breasts. Sprinkle the breasts with salt and pepper on each side.

12. Place the duck breasts skin side down in a cold, dry skillet. Turn the heat to medium and cook until much of the fat has rendered out, about 5 minutes. Increase the heat to medium-high and continue to cook until

Recipe continues on next page

the skin becomes golden brown, thin, and crispy, which should take 5 minutes longer. Turn the breasts over and cook until medium-rare (140°F on an instant-read thermometer), about 3 minutes. Allow the duck breasts to rest for at least 5 minutes before slicing.

MAKE THE CONFIT

13. Preheat the oven to 225°F. Melt the poultry fat in a small saucepan. Briefly rinse the salt off the duck pieces (you don't want to rinse it all away) and pat dry. Arrange the duck pieces in a single layer in a 9- by 13-inch baking dish. Pour the melted fat over the duck (the duck pieces should be covered by fat) and place the confit in the oven. If you don't have enough duck fat, you can augment the fat with poultry fat from another source. You can also augment with olive oil, or even melted lard or tallow.

14. Cook the confit slowly at a very slow simmer — just an occasional bubble — for about 3 hours, until the duck is tender and can be easily pulled from the bone. Remove the confit from the oven.

15. Carefully remove the duck pieces from the fat and transfer to a storage container. Cover and transfer to the refrigerator.

16. Strain the fat and cool in a separate container for at least 8 hours, or overnight. If the fat is in a clear container, you will see that it has separated into fat and jellied juice. Carefully scrape off the fat. Store the juices in the refrigerator or freezer to use as the base for a soup or sauce. Melt the fat and pour it over the refrigerated duck confit pieces. If the duck is completely covered by the fat, it should keep in the refrigerator for at least 1 month (or longer if you live in a French farmhouse). If needed, top off with more poultry fat (preferred) or olive oil. Be sure to completely cover the duck with fat.

SERVING DUCK CONFIT

17. Preheat the oven to 400°F. Wipe off most, but not quite all, of the fat, then put the duck pieces on a sheet pan. Roast for about 15 minutes, until the skin is crisp. Traditional accompaniments are sautéed potatoes (page 162) and a frisée salad dressed with a sharp vinaigrette (page 129). Lentil salad is another classic accompaniment; basically, anything that cuts the richness of the duck works well.

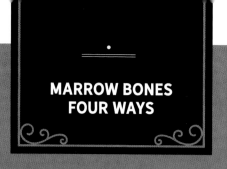

MARROW BONES FOUR WAYS

Marrow bones are rich in flavor and fat. Why not put them to good use? Just roasting the bones renders out usable beef tallow (about 1 tablespoon per pound of bone). Then you can enjoy the marrow on toast or as an herbed butter on toast or steaks, and the bones can be made into broth, with or without the marrow.

ROASTED MARROW BONES

There are those who insist on salting the bones and then blanching them before making anything with them. I don't bother — I'm okay with the beefy, earthy flavor of my bones. The bones can be split horizontally for a more elegant presentation and to make accessing the marrow super easy, but that isn't necessary either. Just preheat the oven to 450°F. Put the bones in a roasting pan, marrow side up, and roast for about 15 minutes. If you are going to enjoy the marrow plain or in a butter, remove from the oven. Pour off the fat and reserve. Serve the marrow bones with toast and allow each diner to scoop out the marrow and spread on toast; special spoons exist for this, but a regular teaspoon or a table knife works fine.

HERBED ROASTED MARROW BONES

This requires marrow bones that are split horizontally. Roast the bones as above for 10 minutes. Meanwhile chop some fresh herbs — parsley, thyme, sage, rosemary — and combine with some flaky sea salt. You can also add an equal amount of breadcrumbs. Sprinkle on the bones after they have roasted for about 10 minutes. Roast for 5 minutes longer. Serve hot with toasts. A sharp parsley, arugula, or frisée salad dressed with a mustardy vinaigrette is a nice accompaniment.

MARROW BUTTER

This butter is rich! It can be dolloped on grilled steaks or spread very lightly on toast. Put 6 garlic cloves, ½ cup fresh parsley, 2 tablespoons fresh sage leaves, 2 tablespoons fresh thyme leaves, and 1 tablespoon fresh rosemary leaves in a food processor and process until finely chopped. Add ½ cup softened butter and the roasted marrow of about 4 pounds of beef marrow bones and process to blend. Season generously with salt. (Makes about 2 cups.) It's a good idea to divide the marrow butter into ¼-cup portions and freeze on a sheet pan. Then wrap in plastic wrap, put the wrapped portions in a bag, and freeze, thawing a small amount at a time as needed.

BONE BROTH

Roast the bones as above. Removing the marrow makes a weaker broth, or one that requires more cooking down, but either way works. Combine the roasted bones with a couple of onions, celery stalks, and parsley sprigs, and cover with water. Bring just to a boil, then reduce the heat and simmer for hours, until the broth has the flavor you want. Refrigerate and use within 5 days or freeze for longer storage. There will be some fat on top of the cooled broth. This fat can be lifted off the broth, scraped clean, and used for stovetop cooking.

BETTER CROUTONS

Nothing perks up a salad or a bowl of soup more than a crunchy crouton. For the thrifty home cook, there's no better use for stale bread. Add duck fat to the mix and you have an above average crouton — but truly any fat works magic here.

Makes 4 cups

- 2 garlic cloves, peeled
- ¼ cup (1.7 ounces/50g) any animal fat
- 1 teaspoon dried oregano, thyme, or Italian seasoning
- 4 cups stale crustless bread cubes (sliced into ¾-inch pieces)
- Salt and freshly ground black pepper

1. Preheat the oven to 350°F. Place a large sheet pan in the oven to preheat at the same time.

2. Grate the garlic with a Microplane or very, very finely mince.

3. Remove the hot sheet pan from the oven and add the fat, turning and tilting the pan until the fat is melted and covers the bottom. Sprinkle with the garlic and herbs. Add the bread cubes and toss the cubes until well coated with the fat. Shake the pan to spread out the cubes, and sprinkle with salt and pepper.

4. Bake for about 15 minutes, turning the bread cubes as needed, until the cubes are golden on all sides.

5. Drain the croutons on paper towels before serving. Leftovers can be stored for about 1 week in an airtight container at room temperature or for up to 6 months in the freezer.

DOUBLE-CRUST PASTRY MADE WITH LARD

Lard pastry handles beautifully and makes flaky, tender piecrusts. Although it is quicker to make the crust in a food processor, cutting the lard into the flour with a pastry cutter or two knives — or by rubbing it in with your fingers (my preferred method) — results in a more tender, flakier crust. If you like, you can swap some of the lard for butter for heightened flavor. A little sugar (1 tablespoon) brightens the flavor of an all-lard crust, but my jury is out on whether two or three tablespoons of sugar is best for a sweet crust. In my opinion, the sweeter the filling, the less sugar is needed in the crust: A tart apple pie benefits from more sugar, while a chocolate cream pie doesn't need it.

Makes enough for 1 double-crust pie or 1 galette

- 2½ cups unbleached all-purpose flour
- 1½ teaspoons fine sea salt
- 1–3 tablespoons sugar
- ¾ cup (5.5 ounces/155g) leaf lard
- ⅔ cup very cold water, plus more as needed

1. Combine the flour and salt in a large bowl. Add 1 tablespoon sugar for a savory pie or 2 to 3 tablespoons for a sweet pie, and whisk until well mixed. Cut in the lard with a pastry cutter or two knives or rub in with your fingertips until the mixture has a pebbly, sandy consistency. Stir in the water until well mixed. Alternatively, combine the flour, salt, and sugar in the bowl of a food processor. Cut the lard into pieces and dot over the top of the flour mixture. Pulse until the mixture is just combined. Add the water and pulse until well mixed. You should be able to form the mixture into a ball. If needed, add more water, a teaspoon at a time, until the dough will form a ball.

2. Gather the dough into a single ball if making a galette or two balls if making a double-crust pie. Wrap well in plastic wrap and refrigerate for about 30 minutes.

TO MAKE AND BAKE
A SINGLE CRUST

Divide the recipe in half. Form the dough into a ball and shape into a flattened disk. Chill for at least 30 minutes. Roll out the dough on a well-floured work surface with a well-floured rolling pin to a thickness of ⅛ inch. Roll the dough onto the rolling pin to transfer it to a 9- or 10-inch pie plate, tucking it into the edges. Trim to leave a 1-inch overhang. Fold under and crimp the edges. Prick the bottom and sides of the dough with a fork at ½-inch intervals. Freeze for about 30 minutes.

Meanwhile, preheat the oven to 425°F with a rack in the lower third of the oven. Remove the unbaked pie shell from the freezer and bake for 20 to 22 minutes, until golden brown. Cool on a rack before filling.

DOUBLE-CRUST PASTRY MADE WITH POULTRY FAT

This piecrust is just as good with a sweet pie as it is with a savory pie. Poultry fat — chicken, duck, or goose — makes a flaky, crisp piecrust. The crust remains surprisingly crisp even if the pie hangs around for a few days. The dough is somewhat more fragile than a pie dough made with lard or tallow but is still pretty easy to handle, provided your work surface and rolling pin are very well floured. It works best when the fat is chilled before combining, the dough is chilled before rolling out, and the pie shell is chilled before filling.

Makes enough for 1 double-crust pie or 1 galette

- 2½ cups unbleached all-purpose flour
- 1½ teaspoons fine sea salt
- 1–3 tablespoons sugar
- ¾ cup (5.5 ounces/155g) any poultry fat
- ½ cup very cold water, plus more as needed

1. Combine the flour and salt in a large bowl. Add 1 tablespoon sugar for a savory pie and 2 to 3 tablespoons for a sweet pie, and whisk until well mixed. Add the fat and stir with a fork until the mixture has a pebbly, sandy consistency. Drizzle the water over the flour mixture and stir with a fork until well mixed. You should be able to form the mixture into a ball. If needed, add more water, a teaspoon at a time, until the dough will form a ball.

2. Gather the dough into a single ball if making a galette or two balls if making a double-crust pie. Wrap well in plastic wrap and refrigerate for about 30 minutes.

DOUBLE-CRUST PASTRY MADE WITH TALLOW

Tallow is hard, and little shards created by grating or finely chopping will make tears in the pastry as it is rolled out. Therefore, I recommend grinding it with some flour to get the tallow into very fine pieces. Don't worry, your piecrust will still be flaky, even though the fat is evenly dispersed and quite fine. A tallow piecrust has a slightly freckled appearance.

Makes enough for 1 double-crust pie or 1 galette

- 7 ounces/200g tallow
- 2½ cups unbleached all-purpose flour
- 1½ teaspoons fine sea salt
- 1–3 tablespoons sugar
- ⅔ cup ice water, plus more as needed

1. Combine the tallow and ½ cup of the flour in a food processor and process until it has the texture of tiny pebbles. Add the remaining 2 cups flour, the salt, and sugar, using 1 tablespoon sugar for a savory pie and 2 to 3 tablespoons for a sweet one. Pulse the mixture until just combined. Add the water and pulse until well mixed. You should

be able to form the mixture into a ball. If needed, add more water, a teaspoon at a time, until the dough will form a ball.

2. Gather the dough into a ball, flatten into a disk, wrap well in plastic wrap, and refrigerate for at least 30 minutes, or up to 1 day. Allow to soften at room temperature for a few minutes if it has chilled for more than 30 minutes.

PASTY PASTRY

The original Cornish pasties were made with a crust tough enough to withstand a trip down into the mine in a miner's coat pocket. The pasty had to have a seam that functioned as a handle for the pastry. Miners, who had no way to wash their hands at lunchtime, could hold the pasty by the handle and then discard it after eating the rest. This pasty crust needs to be just sturdy enough to fill and bake without leaking.

Makes enough for 10–12 pasties

- 4 cups unbleached all-purpose flour
- 2 teaspoons fine sea salt
- 1 cup (7 ounces/200g) any animal fat (except bacon grease)
- 2 eggs
- ¾ cup ice water, plus more as needed

1. Whisk together the flour and salt in a large bowl.

2. If you are making the pastry with tallow, put the tallow in a food processor along with ¼ cup of the flour mixture. Grind until the mixture resembles tiny pebbles. Add the eggs and pulse until mixed in. Add the water and pulse until the mixture can be pressed together into a ball, adding more water if needed.

If you are working with lard, break up the lard with your fingers and toss the pieces in the flour mixture to coat them thoroughly with flour. Rub the fat into the flour until the mixture becomes pebbly using your fingertips, two knives, or a pastry blender. Beat the eggs with the ice water. Make a well in the flour and add about three-quarters of the egg mixture. Mix the dough by hand or with a fork, just until all the flour is moistened, adding additional egg mixture by the tablespoon if needed. Gather the dough together into a ball.

If you are working with a poultry fat, stir the fat into the flour until the mixture becomes pebbly using your fingertips, two knives, or a pastry blender. Beat the eggs with the ice water. Make a well in the flour and add about three-quarters of the egg mixture. Mix the dough by hand or with a fork, just until all the flour is moistened, adding additional egg mixture by the tablespoon if needed. Gather the dough together into a ball.

Recipe continues on next page

Pasty Pastry, continued

3. Transfer the dough to a lightly floured surface and knead lightly until fairly smooth (the dough will have a pebbled appearance if made with tallow or lard). Divide into two pieces and shape into logs. Wrap each in plastic wrap and refrigerate for at least 1 hour, or up to 6 hours.

4. Roll, fill, and bake according to the recipe directions.

SWEET TART PASTRY MADE WITH LARD OR POULTRY FAT

This sturdy pastry is more cookie dough than pie dough and it is patted into place, which makes it super easy to make.

Makes enough for one 9-inch tart

- 1½ cups unbleached all-purpose flour
- ½ cup confectioners' sugar
- 1 teaspoon fine sea salt
- ½ cup (3.5 ounces/100g) leaf lard or poultry fat
- 1 egg, beaten

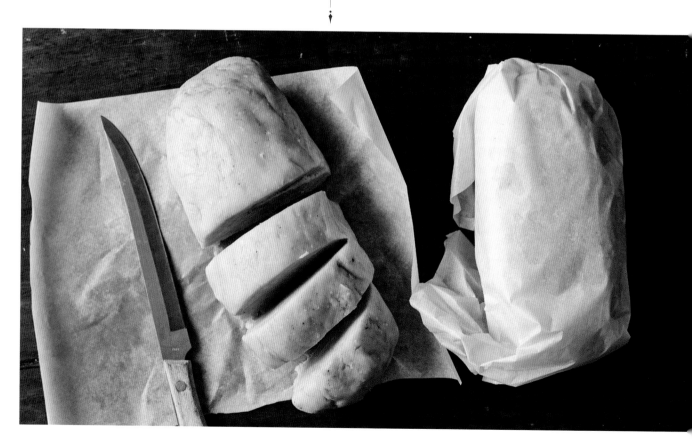

1. Combine the flour, confectioners' sugar, and salt in the bowl of a food processor. Cut the lard into pieces and distribute the pieces over the dry ingredients, or drizzle the poultry fat over the flour mixture. Pulse until the fat is coarsely cut in and the mixture resembles pebbly, coarse sand. With the motor running, slowly pour in the egg.

2. Grease a 9-inch tart pan with a removable bottom. Dump the mixture into the pan and press evenly across the bottom and up the sides of the tart pan, pressing just hard enough to make a solid surface, but not so hard that the pastry loses its crumbly texture. Prick all over with a fork.

3. Freeze the crust for 30 minutes to 1 hour before baking.

4. To fully or partially bake the crust, preheat the oven to 375°F. Put the tart pan on a sheet pan and bake the crust for 25 minutes to partially bake, or about 30 minutes to fully bake, until firm and golden brown.

SWEET TART PASTRY MADE WITH TALLOW

This pastry has the crisp texture of a shortbread cookie and is absolutely easy to make because it doesn't require any rolling out.

Makes enough for one 9- or 10-inch tart

- 3.5 ounces/100g tallow, chopped
- 1½ cups unbleached all-purpose flour
- ½ cup confectioners' sugar
- 1½ teaspoons fine sea salt
- 1 egg, beaten

1. Combine the tallow with ½ cup of the flour in the bowl of a food processor. Process until the mixture forms even little pellets. Add the remaining 1 cup flour, the confectioners' sugar, and salt. Pulse until the mixture looks like coarse sand. With the motor running, pour in the egg slowly.

2. Grease a 9- or 10-inch tart pan with a removable bottom. Dump the mixture into the tart pan and press it evenly across the bottom and up the sides of the tart pan, pressing just hard enough to make a solid surface, but not so hard that the pastry loses its crumbly texture. Prick all over with a fork.

3. Refrigerate or freeze the crust for 30 minutes to 1 hour before baking.

4. To fully or partially bake the crust, preheat the oven to 375°F. Place the tart pan on a sheet pan. Bake the crust for 25 minutes to partially bake, or about 30 minutes to fully bake, until firm and golden brown.

METRIC CONVERSIONS

FAT VOLUME TO WEIGHT CONVERSIONS

1 tablespoon	0.45 ounce	12.8g
2 tablespoons	0.8 ounce	25g
3 tablespoons	1.4 ounces	40g
4 tablespoons	1.7 ounces	50g
5 tablespoons	2 ounces	60g
⅓ cup	2.3 ounces	65g
6 tablespoons	2.6 ounces	75g
7 tablespoons	3 ounces	90g
½ cup	3.5 ounces	100g
⅔ cup	4.4 ounces	125g
¾ cup	5.5 ounces	155g
1 cup	7 ounces	200g

TEMPERATURE

TO CONVERT	TO	
Fahrenheit	Celsius	subtract 32 from Fahrenheit temperature, multiply by 5, then divide by 9

COMMON OVEN TEMPERATURES

US	METRIC
200°F	90°C
225°F	110°C
250°F	120°C
275°F	140°C
300°F	150°C
325°F	160°C
350°F	180°C
375°F	190°C
400°F	200°C
425°F	220°C
450°F	230°C
475°F	250°C
500°F	260°C
550°F	290°C (broiling)

RESOURCES

Food writers often call for specific foods to be used in their recipes, telling the reader to find them at a certain type of ethnic market or specialty food store — or your local butcher. The very phrase "local butcher" can set people off, because local butchers don't exist everywhere. Still, the search term "local butcher" does yield numerous listings of small shops that sell locally raised meats. Those local butcher shops are one of the first places to go when looking for sources of pasture-raised meats and fats. Farmers' markets are also excellent sources.

If you search for "lard" online, you will see many, many sources for rendered lard — some not organic or pasture-raised or even purely lard, so be careful of your sources. Many of the sources for lard are sold on Etsy, the online handmade-goods retailer. Those sellers are bound by the laws of their states, but it is unclear who verifies or inspects their products. You do not want any product that lists on its label "hydrogenated" or "BHA, propyl gallate, and citric acid added to help protect flavor" or "no refrigeration." I have listed some online sources that I have found dependable and reasonably priced. The market is booming right now, and more sources are appearing in both local and national marketplaces.

PRODUCTS AND INFORMATION

D'ARTAGNAN

A reliable (if expensive) source of "humanely raised" poultry and poultry fat. The website contains lots of helpful information and solid, if upscale, recipes. D'Artagnan's specialty meats (especially duck) and duck fat are pretty widely distributed to specialty food stores.
www.dartagnan.com

EATWILD

Founded in 2001, Eatwild's mission is to promote the benefits of choosing meat, eggs, and dairy products from 100 percent grass-fed animals or other nonruminant animals fed their natural diets. Eatwild features a state-by-state directory of local farmers who sell directly to consumers. Plus it has a wealth of well-documented information on the health benefits of consuming grass- and pasture-raised meats and fats and the environmental benefits of pasture-raising animals.
www.eatwild.com

FATWORKS

Curious about the taste of bison fat? Looking for pasture-raised goose fat that isn't imported from France? Fatworks has it all, and more. They even sell 1-ounce jars for sampling if you aren't sure which type of fat you'd like to cook with. They periodically feature the farms they source from and also provide lots of recipes and information about cooking with animal fats. Their fat is pretty widely distributed, and they offer a guide to stores in the United States that carry their products, to save you the cost of shipping.
fatworksfoods.com

TENDERGRASS FARMS

Another reliable source of organic, pasture-raised, rendered lard.
tendergrass.com

US WELLNESS MEATS

Since 2000, US Wellness Meats has been selling meats and meat products from farmers who are committed to returning to the environmentally sound rotation grazing practices that are best for their sheep, cattle, pigs, and poultry. They sell rendered lard, tallow, and duck fat, as well as unrendered suet from lamb and beef.
grasslandbeef.com

BOOKS AND ARTICLES

McLagan, Jennifer. *Fat: An Appreciation of a Misunderstood Ingredient, with Recipes.* Ten Speed Press, 2008.

Morell, Sally Fallon. *Nourishing Fats: Why We Need Animal Fats for Health and Happiness.* Grand Central Life & Style, 2017.

Nestle, Marion. *Food Politics* (blog). www.foodpolitics.com.

Ramsey, Drew, and Tyler Graham. *The Happiness Diet: A Nutritional Prescription for a Sharp Brain, Balanced Mood, and Lean, Energized Body.* Rodale Books, 2011.

Ruhlman, Michael. *The Book of Schmaltz: Love Story to a Forgotten Fat.* Little, Brown & Co., 2013.

Ruhlman, Michael, and Brian Polcyn. *Charcuterie: The Craft of Salting, Smoking, and Curing.* W. W. Norton, 2005.

Taubes, Gary. "The Soft Science of Dietary Fat," *Science* 291, no. 5513 (March 30, 2001): 2536–45.

Teicholz, Nina. *The Big Fat Surprise: Why Butter, Meat, and Cheese Belong in a Healthy Diet.* Simon & Schuster, 2014.

Index

Page numbers in *italic* indicate photos; pages numbers in **bold** indicate charts.

KEEP YOUR CREATIVITY COOKING

WITH MORE BOOKS BY ANDREA CHESMAN

Preserve your own home-grown feast with 150 delicious recipes for salsa, relishes, fermented pickles, chutneys, and much more. Expert advice and simple recipes will have you putting up everything from apples to zucchini in no time.

This collection of 175 recipes makes your garden-fresh vegetables shine. In addition, 14 simple master recipes show you how to easily accommodate whatever might be especially bountiful in your produce basket.

Enjoy local produce year-round! This collection of 270 recipes — starring jewel-toned root vegetables, hardy greens, sweet winter squashes, and potatoes of every kind — brings a variety of delightful flavors and an impressive array of nutrients to your winter table.

31901064532593